CITIZEN SCIENCE IN THE DIGITAL AGE

T0288161

RHETORIC, CULTURE, AND SOCIAL CRITIQUE

CITIZEN SCIENCE

IN THE DIGITAL AGE

RHETORIC, SCIENCE, AND PUBLIC ENGAGEMENT

JAMES WYNN

THE UNIVERSITY OF ALABAMA PRESS   TUSCALOOSA

The University of Alabama Press
Tuscaloosa, Alabama 35487-0380
uapress.ua.edu

Hardcover edition published 2017.
Paperback edition published 2019.
eBook edition published 2017.

Inquiries about reproducing material from this work should be
addressed to the University of Alabama Press.

Typeface: Scala and Scala Sans

Cover design: Michael West

Paperback ISBN: 978-0-8173-5952-2

A previous edition of this book has been catalogued by the Library
of Congress as follows:

Library of Congress Cataloging-in-Publication Data

Names: Wynn, James, 1972–
Title: Citizen science in the digital age : rhetoric, science, and public
    engagement / James Wynn.
Description: Tuscaloosa : The University of Alabama Press,
    [2017] | Series: Rhetoric, culture, and social critique | Includes
    bibliographical references and index.
Identifiers: LCCN 2016023996| ISBN 9780817319434 (cloth : alk.
    paper) | ISBN 9780817390860 (e-book)
Subjects: LCSH: Science—Public opinion. | Mass media and
    technology. | Technology—Social aspects. | Digital media—
    Social aspects.
Classification: LCC Q175.5 .W96 2017 | DDC 303.48/3—dc23
LC record available at https://lccn.loc.gov/2016023996

Contents

Illustrations

Acknowledgments

The adage "it takes a village to raise a child" I think can be reasonably applied to books. It was only with the generous help of institutional funding, interviewees, and colleagues that this project has come to be in its current form. In developing this project, for example, visuals emerged as an important locus for exploring the impact of digital technologies and citizen science on communication and argument. In chapter 2, I make the case that the Internet and Internet connectable devices have allowed for the emergence of new kinds of risk representations. To arrive at this conclusion, I rely on evidence from visuals. Although their existence in the book seems like an unremarkable fact of scholarship, I almost did not include them because of the high cost of making the visuals suitable for reproduction. Their presence in this book is made possible through a generous grant from Carnegie Mellon's Dietrich College Berkman Faculty Development Fund.

In addition to material support, the scholarship in this book would not have been possible without the cooperation of individuals and institutions that were willing to share their insights and creative productions with me. I would start by thanking the developers of RDTN and Safecast whose work I write about in chapters 2 and 3. In particular, I would like to express my gratitude to the members of these organizations in Japan and on the West Coast who got up early or stayed up late to be interviewed about their work. They include Akiba, Sean Bonner, Peter Franken, Marcelino Alvarez, and Azby Brown. These interviewees provided behind-the-scenes details about the founding and development of RDTN and Safecast, access to their Google Groups discussion board, and permission to use visuals of the Safecast map in this book. Next, I would like to express my appreciation to Anthony Watts and Roger Pielke Sr. whose Surface Stations project is the subject of chapter 4. Both gentlemen offered important historical background on the project and insight into the goals and challenges of its development. I am also thankful to Muki Haklay and Louise Francis of the University College of London who supplied me with details about their citizen-science sound-mapping project described in chapter 5. They shared with me a transcript and audio file of the community meeting in which citizen scientists and borough representatives discussed the results of their sound-mapping project. The details of community interactions made accessible by these sources provide crucial support for the arguments in chapter 5. Finally, I would like to add a special thanks to Annie Griffin whose citizen-science sound-mapping project did not make it into this book but who generously agreed to be interviewed at a time when I thought it would.

Last, but not least, I want to thank all of those people who have given their intellectual and emotional support to help me develop this book. Among the former, I would like to recognize my graduate research assistant Christy Gelling whose incredible research skills were responsible for identifying many of the texts used as evidence in chapter 4. I also owe her a debt of gratitude for suggesting that the Surface Stations project might be worth exploring. Without her keen sense of scholarly potential, this chapter might have been on something entirely different. I would also like to express my debt to my rhetoric of science colleagues Lynda Walsh and Lisa Keränen who generously agreed to read parts of this book and give feedback. Their insights helped me strengthen my argument and make it more accessible. The clarity and strength of the book's arguments are also thanks to all of my colleagues in the field who have attended the talks I have given on chapters in the book and offered insights and critiques. Finally, I would like to thank my wife and my kids for their emotional support and patience as I worked on this project. Without this village, this book would not have been possible.

CITIZEN SCIENCE IN THE DIGITAL AGE

Introduction

In the deep jungle, a group of pygmies out for a hunt stop and gather around the shattered remains of trees strewn across the jungle floor. One of the hunters pulls out a small GPS device and marks the spot. Thousands of miles away in an office at the University College London a researcher adds the coordinates to an online map of the Congo already speckled with reported signs of illegal logging activity. Across the Atlantic in Seattle, a teenager plays a computer game carefully manipulating the branch-like shapes of proteins watching her score rise and fall as she optimizes their configurations. Across town, researchers at the University of Washington harness her folding skills to identify protein configurations that might help them synthesize more effective drugs to fight AIDS, cancer, or Alzheimer's.

These brief vignettes add to an already substantial list of illustrations of a truism of the digital age: that the Internet has changed the way we communicate and interact with one another. The ubiquity of digital technology and the myriad ways that it has become integrated into the fabric of modern life have changed the quantity of available information, the speed with which it can be accessed, and the distances across which it can be broadcast. Though there is a widely shared recognition *that* the Internet has transformed the ways we communicate and interact, there is less frequently an understanding of *how* this technology actually impacts our communications and interactions. Are the pygmies benefitting socially, economically, or politically by electronically marking illegal logging sites? Do virtual protein folding games help laypeople understand the science they are helping advance? These questions are important for researchers interested in tracking the development of the emerging phenomena known as "citizen science."

For more than a decade, "citizen science" has been receiving increased attention because of its potential as a cost-effective method of gathering massive data sets and as a means of bridging the intellectual divide between laypersons and scientists. Despite its possible material and social benefits, little effort has been devoted to understanding whether, and if so in what ways, citizen science has been shaping interactions between laypersons, science, scientists, and policymakers. This book explores these poorly understood dimensions of citizen science by assessing real-world discourse, textual and visual, generated by it. In doing so, it endeavors to expand the scholarly investigations in the field of rhetoric of science on a variety of subjects like visual communication, argument from expertise, identification between scientists and laypersons, and the interaction of scientific and nonscientific *logoi* in policy argument.

Citizen Science

Before discussing the ways in which citizen science might expand our understanding of science and argument, it is important to consider what kinds of activities the phrase "citizen science" references and what dimensions of these activities have or have not attracted the attention of researchers. In its most generic sense, "citizen science" refers to the emerging practice of using digital technologies to crowdsource information about natural phenomena. This basic framing is used in many of the writings of scientists and scientific popularizers who discuss citizen science. In the recently published *Citizen Science: Public Participation in Environmental Research*, for example, John Fitzpatrick, director of the Cornell Lab of Ornithology, explains, "as currently configured, [citizen-science] projects . . . engage participants exclusively as sensors to record and transmit observations [that] were designed by scientists and organized to gather data that can be analyzed at the back end," (Fitzpatrick 238). In a similar vein, science journalist Jeffrey Cohn explains in the popular journal *Bioscience*, "The term 'citizen scientists' refers to volunteers who participate as field assistants in scientific studies. . . . Typically, [they are] volunteers [who] do not analyze data or write scientific papers, but they are essential to gathering the information on which studies are based" (Cohn 193).

These general descriptions of citizen science situate the activity in a century-and-a-half-long tradition of citizens volunteering for governments and scientific institutions to gather information about natural phenomena like weather and astronomical events.[1] Though modern citizen science certainly shares some of the basic features of its predecessors, it is only its recent digital form that has been formally recognized as a scientific activity. In fact, according to the *Oxford English Dictionary* the phrase "citizen science" did not appear in print in English until 1989. Scientific journal databases also show that citizen science was not a topic of discussion amongst scientists until the twenty-first century. Ecologist Jonathan Silvertown explains, "Despite its deep roots, recognition that the modern form of citizen science is a distinct activity with its own constituency of practitioners is recent. In January 2009, the *ISI Web of Knowledge* database contained only 56 articles explicitly dealing with 'citizen science'. . . . Nearly all the articles (80%) listed in the *Web of Knowledge* had been published in the last five years" (470).

According to Silvertown's calculations using the *Web of Knowledge*, the distinct enterprise known as citizen science emerged in some substantial way between 2004 and 2009.[2] But what precisely sets this new practice apart from its predecessors? Elaborations on the practice in the scientific literature suggest that some scientists make a distinction on quantitative grounds. For instance, Cooper and coauthors write in a 2007 article "Citizen Science as a Tool for Conservation in Residential Ecosystems": "The *Citizen Science* model engages a dispersed network of volunteers to assist in professional research using methodologies that have been developed by or in collaboration with professional researchers. The public plays a role in data collection across broad geographic regions. . . . The use of dispersed participants in citizen science creates the capacity for research at a broadly ambitious scale in contrast to localized volunteer-based research projects" (2).

Cooper and her coauthors' comparison of citizen science to other kinds of volunteer research highlights the fact that it involves more people spread out over a larger geographic area gathering more data. However, for other scientists the difference is not simply in the number of lay participants involved or the amount of data they collect but also in the means by which these new modes of data collection are made possible. Unlike volunteer projects from previous periods, modern citizen science is unique because it capitalizes on the existence of widespread computing resources and the Internet that connects them. In fact, the emergence of citizen science as a new category of scientific activity tracks closely with the development and expansion of these technologies. In a 2010 retrospective examining the emergence of citizen science, "Citizen Science as an Ecological Research Tool: Challenges and Benefits," authors Dickinson, Zuckerberg, and Bonter mark the relationship between the two: "Citizen science projects have proliferated in the past decade, with the ability to track the ecological and social impacts of large scale environmental change through the Internet. Sophisticated Internet applications effectively utilize crowdsourcing for data collection over large geographic regions, offering opportunities for participants to provide, gain access to, and make meaning of their collective data" (150).

As Dickinson, Zuckerberg, and Bonter explain, the rise of the Internet has allowed scientists to effectively tap into new resources and provided new opportunities for lay volunteers to participate in science. A brief look at some examples of modern citizen science provides a sense of the ways in which laypeople and scientists can now work together and how these new relationships might be influencing the material and social conditions of science.

Potential Material and Social Benefits of Citizen Science

Modern science is a materially intense enterprise requiring massive computing power and human resources to both collect and process data. In the face
of stretched and tightening budgets, scientists have recognized that citizen
science promises to deliver resources they need to continue advancing their
research and have developed some remarkably savvy ways of harnessing the
material and intellectual resources of the public to advance science. One of
the earliest citizen-science projects SETI@home (1999), for instance, takes
advantage of private citizens' computing resources by asking them to allow
researchers working on the Search for Extra Terrestrial Intelligence (SETI)
to borrow their home computers to process data from their telescope arrays
("About SETI@home"). Although a single laptop or desktop hardly matches
the computing power of ultrafast supercomputers, SETI's capacity to out-
source processing to thousands of privately owned devices has provided it
with a substantive low-cost boost to their processing power.

In addition to using citizens' personal computers as resources for processing scientific data, scientists have also employed digital technology to
tap into their cognitive abilities to solve scientific puzzles. In 2008, for example, researchers at the University of Washington developed the game FoldIt
and invited the Internet-using public to help them figure out how important
classes of proteins fold themselves. By solving online folding puzzles, game
participants helped scientists develop protein structure models that could
be fabricated and tested in the lab ("FoldIt," Wikipedia). The puzzle-solving
abilities of FoldIt players were so good that they made a critical contribution
to AIDS research with their discovery of the configuration of the retroviral
protease M-PMV, which is a key protein in simian AIDS (Khatib et al. 3).

While FoldIt draws on laypeople's puzzle-solving skills to advance scientific research, digital citizen-science projects like the Feeder Watch and Galaxy Zoo use the Internet to enlist an army of public participants to gather
and process data. Data-hungry fields like ornithology, phenology,[3] and ecology
rely on geographically distributed volunteers to collect data on phenomena
like bird migration and the blooming periods of plants. Given the distributed nature of these events, scientists recognize the impossibility of having
enough researchers and graduate students to cover them. They are, therefore,
keenly interested in using citizen-science projects like Feeder Watch to enlist
the help of lay volunteers to gather data. Conversely, researchers in data-rich
fields like astronomy and zoology see citizen scientists as resources for processing mountains of data that have already been collected by tapping into
their ability to categorize natural phenomena. In the Galaxy Zoo astronomy
project, for instance, lay volunteers are asked to sift through pictures of far-
off galaxies and classify them by type. Without their help, it would be im-

possible for scientists to work through the backlog of data their telescopes have collected or have any hope of processing the ever-increasing volume of data that pours in every year.

As these examples illustrate, the spread of networked computing has made working with the public on scientific research increasingly more feasible and attractive. As a consequence of these new technologies, scientists have been able to harness the computing power, puzzle-solving skills, and data-gathering and processing capabilities of the public in ways that were previously unimaginable. By helping with these tasks, citizen science keeps massive scientific projects moving forward in the face of the increasing material challenges of doing science.

Many scientific and popular discourses about citizen science focus on its benefits in material terms; however, in some instances its potential social payoffs have been the center of attention. In *Citizen Science: Public Participation in Environmental Research*, for instance, editors Janis Dickinson and Rick Bonney suggest that as digital-age technology creates more opportunities for laypeople and scientists to work together, scientists will have increased opportunities to reevaluate citizen expertise as well as their own status as experts: "Although scientists who engage with the public may initially buy into the merits of combining education and research for the public good, their involvement in citizen science may cause them to challenge the deficit model and other preconceptions they have about participants. In this regard, citizen science is fertile ground for change in how scientists view the public and, ultimately, how scientists view themselves" (Dickinson and Bonney 11).

Whereas Dickinson and Bonney recognize that citizen science might influence scientists' perceptions about lay knowledge, they and other scientists also acknowledge that it can provide opportunities for educating the public about science and encouraging scientific perspectives on significant public issues. In discussions about next-generation citizen-science projects, for example, scientists have recommended devoting more energy to involving citizens in the development of research questions and experimental design (Bonney et al., *Public Participation* 48–49). By crossing the socio-epistemic boundary between researchers and volunteer data collectors, these efforts are aimed at promoting public identification with scientific practices and perspectives. This increased identification, researchers hope, will slow what they see as steady erosion of the public's critical faculties. Dickinson and Bonney explain, "Increasing the public's ability to think critically about science is perhaps the only way to diminish the impact of information put forth by corporate entertainment news sources, which are rife with misrepresentations of important scientific and political issues" (Dickinson and Bonney 11).

From a scientific perspective, citizen science has the potential to change the social conditions of science by improving scientists' understanding of lay expertise, gained from personal experience with nature, and laypeople's understanding of scientific practices and perspectives. In public discussions of citizen science, even broader social benefits are imagined, including the political empowerment of laypeople participating in citizen science and the epistemic democratization of science. Writing in *Nature* about the capacity of citizen science to politically empower laypeople, for example, journalist Katherine Rowland explains, "The next generation of citizen science attempts to make communities active stakeholders in research that affects them, and uses their work to push forward policy progress" (Rowland par. 2). In her article, Rowland cites citizen-science programs that provide Congolese pygmies with GPS tracking devices to identify illegal logging as well as supply Londoners with recording devices to track neighborhood noise and air pollution as instances of political empowerment projects.

Whereas Rowland frames the benefits of citizen science in terms of political empowerment, the *Boston Globe*'s Gareth Cook focuses on its capacity to erode the intellectual barriers between scientists and laypersons. Using websites like FoldIt and Galaxy Zoo as examples, he argues, "Science is driven forward by discovery, and we appear to stand at the beginning of a democratization of discovery" (Cook, "How Crowdsourcing"). He explains that as the public becomes more involved with data gathering and assessment, scientists will have to recognize the value of lay contributions to the advancement of knowledge.

Is Citizen Science Materially and Socially Transformative?

Both popular and scientific sources highlight the potential of citizen science to transform the material conditions of science as well as the social, political, and epistemological relationships between laypersons, scientists, and policymakers. Though these sources discuss optimistically its transformative potential, we should inquire, "Is there evidence to suggest that these transformations are happening?" A number of discussions in the scientific and technical literature suggest that the capacity of citizen science to transform the material conditions for doing science is quite real. In an article assessing the costs and benefits of the Cornell Lab of Ornithology's (CLO) citizen-science projects, researchers from the lab remark, "CLO's current citizen science budget exceeds $1 million each year" to pay for staff, participant support, and data collection, analysis, and curation (Bonney et al., "Citizen Science" 983). Despite these substantial costs, they argue, "considering the high quality of the data that citizen science projects are able to collect . . . the citizen science model is cost effective over the long term" (983). Research on the economics of the SETI@home project provides even more compelling evidence of citizen science's capacity to make big scientific projects materially feasible. In an analysis of the economic benefit of SETI@home, Microsoft analyst Jim Gray described citizen science as "a very good deal" in which "the SETI@home peers donated a billion dollars' worth of 'free' CPU time and also donated 10^{12} watt hours [of electricity] . . . [at a cost of] $100 million" to aid in the search for extraterrestrial life ("Distributed Computing Economics" 3).

Though the material impact of citizen science on scientific projects has been researched and documented, its influence on the social, political, and epistemological dimensions of science has yet to be fully considered. The general absence of research on this subject is documented in a comprehensive report on models of public participation in science from the Center for the Advancement of Informal Science Education (CAISE). In the final pages of the report, the authors explain that understanding how collaborative research projects—including citizen science—influence social engagement and interactions between scientists and laypersons is a major research gap that needs to be filled:

> Understanding the impacts of PPSR [Public Participation in Scientific Research] could push the boundaries of what we currently define as learning in the realm of science, including learning that affects participants' lives in a very broad sense.
>
> Equally important is examining the learning impacts for scientists. What are they learning in terms of science knowledge, science process, and attitude about science? This type of research has not been done at all as far as we can determine. (Bonney et al., *Public Participation* 50)

The research space opened by the lack of assessment of the social outcomes of citizen science provides an opportunity for examining how this emerging phenomenon of the digital age is shaping the relationships between laypersons, science, scientists, and policymakers. Because this investigative space includes social, political, and epistemological dimensions, it invites exploration by fields broadly interested in these phenomena and with the available concepts and methods for describing them. The field of rhetoric has these qualifications. Rhetoric is a discipline that studies persuasive argument assessing choices of language, arrangement, style, and argument as well as the audiences for which and contexts in which these choices are made. Rhetoric's focus not only makes it an important disciplinary starting point for studying citizen science but also marks it as an intellectual space whose conversations might be enriched by examining the historical, material, linguistic, and social dimensions of citizen science. In particular, the subdiscipline of rhetoric of science could most profit from and contribute to this exploration. It is in the scholarship of this area of rhetorical study that this book locates its conceptual center and orients its scholarly contribution.

CITIZEN SCIENCE, RHETORIC OF SCIENCE, AND SCIENCE COMMUNICATION

The field of rhetoric of science focuses attention on how persuasive argument in and about science can be used to advance the perspectives and interests of laypersons, scientists, and policymakers under a variety of social, political, and material conditions. By examining citizen science, this book draws on and contributes to a variety of ongoing discussions in rhetoric of science, including investigations of visual communication, argument from expertise, identification between scientists and laypersons, and the interaction of expert techno-scientific *logos* and nonexpert *logos* in policy argument.

To provide a context for these discussions, chapter 1 explores historical instances of citizen science examining the continuities and differences between data-gathering enterprises in the pre- and post-digital era. By comparing nineteenth- and twentieth-century citizen-science projects with their digital-age counterparts, chapter 1 suggests that though there are similarities in the challenges project developers face, modern digital tools allow for solutions that make citizen science more attractive and important to mainstream institutional science. As the other chapters in this book reveal, these changes in the status and practice of citizen science have impacted interactions between laypersons, science, scientists, and policy makers and created new research spaces for rhetorical scholars to explore.

In the last decade, rhetoric of science scholars have devoted increasingly more attention to the role of visuals in science. In some instances, these investigations have been dedicated to describing the historical development of visuals as a medium for explanation and persuasion (Gross, Harmon, and Reidy 2002; Gross 2009). In others, they have focused on how scientific visuals might be misused by laypersons to make misguided or even misleading claims about natural phenomena (Gibbons 2007). The exploration of citizen science provides a space for expanding the discussion of the role of visuals in communicating techno-scientific information to the public by examining whether, and if so how, laypersons might use the technological affordances of the Internet to create their own visual representations of risk. This subject is taken up in chapter 2, "Reimaging Risk," which looks in detail at the efforts of the grassroots citizen-science group Safecast to create its own visual representations of radiation risk following the Fukushima accident. The chapter explores the questions: Are public representations of risk created by grassroots citizen science different from risk representations in the mainstream media? and if so, What might these differences reveal about grassroots perspectives on risk communication? The chapter pursues these questions by comparing Safecast's Internet-enabled risk visualizations with radiation risk visuals in the mainstream media—the only source where visual representations of risk were broadly disseminated in the public sphere—following the nuclear accidents at Three Mile Island, Chernobyl, and Fukushima.

In addition to expanding the investigation of visual communication, citizen science also opens up new areas for exploring the role of expertise in argument over techno-scientific issues. The issue of expertise and its influence on argument has been considered from a variety of perspectives. Some scholars have focused on the privileging of scientists and scientific evidence and argument in public debates (Katz and Miller 1996; Grabill and Simmons 1998; Simmons 2007). Others have devoted attention to the efforts of laypersons to become experts through the study of scientific literature and scientific reevaluation of lay expertise (Fabj and Sobnosky 1995; Endres 2009). By focusing on the outcomes of citizen-science projects, this book extends the investigation of how laypersons develop their expertise by considering the question, How might changes in *technologies for data gathering* impact the capacity of laypersons to argue as experts? This question is taken up in chapter 3, "Information for and by the People: The Internet and the Rise of Citizen Expertise," which examines Safecast's struggle to defend their citizen-science project against institutional criticism of its reliability. The chapter reveals how Safecast's capacity to invent Internet-connectable devices helped them expand their ability to make technical ethical arguments. The power of data-gathering practices and technologies to affect argument suggests that in some ways the obstacles of expertise—namely the lack of participation in scientific culture and practice—are being eroded by digital-age technologies.

In addition to expertise, rhetoric of science scholars have also recognized the significance that interpersonal understanding and identification can have on argument and persuasion in interactions between laypeople and scientists. Researchers studying scientific involvement in publicly sensitive topics like AIDS, nuclear power, and biohazards, for example, have illustrated how interactions between scientists and laypersons strongly influence beliefs and values in both of these groups and how they influence interactions between them (Fabj and Sobnosky 1995; Kinsella 2004; Waddell 1996). In their discussion of interactions between AIDS activists and doctors, for instance, Valeria Fabj and Matthew Sobnosky note how open communication between these groups often led to improved mutual understandings and the recognition that representatives on either side of the table had the best interest of AIDS sufferers in mind (172). Though scholarly conversations like this one illuminate the mutual positive benefits of closer interactions for laypersons and scientists, they don't substantially entertain the possibility of negative consequences of these interactions or the factors that might contribute to less-than-favorable outcomes.

Chapter 4 of this book, "Warming Relations? The Benefits and Challenges of Promoting Understanding and Identification with Citizen Science," examines the challenges that arise when citizens and scientists work together to address a controversial public issue. To understand these challenges, it explores a collaborative citizen-science project cocreated by a climate scientist, Dr. Roger Pielke Sr., and a climate change skeptic, Anthony Watts, whose purpose was to investigate the site conditions at temperature measurement stations across the US Historical Climatology Network (USHCN).[4] With the help of this case, the chapter explores the questions: Can citizen science be an effective means of promoting identification and mutual understanding between laypersons and scientists? and Can citizen science provide access for laypersons to technical sphere argumentation? A detailed assessment of the project's development reveals that though it emerged organically from a shared interest in the problems of temperature measurement, its capacity to promote identification and mutual understanding were mixed. These findings suggest that citizen science is subject to and shaped by the social political context in which it operates. It therefore requires more than just mutual interest and active collaboration to promote improved identification and understanding between citizens and scientists. It also demands a rhetorical awareness of the possible socio-political and epistemological obstacles to these desired outcomes.

The final subject of this book is the *interaction* between expert *logos* and nonexpert *logos* in techno-scientific debates, which is perhaps the least discussed topic in the rhetoric and rhetoric of science literature. The current scholarship on *logos* has tended to treat expert and nonexpert *logos* separately and has devoted its energy either to defining nonexpert *logos* or justifying its legitimacy in public arguments (Katz and Miller 1996; Grabill and Simmons 1998; Fisher 1987; Fischer 2000). The only scholarly treatment that examines in detail the interaction between expert and nonexpert *logoi* in public debate is Craig Waddell's investigation of the use of mixed appeals in arguments over water quality standards for the Great Lakes. In his arguments, Waddell suggests that successful appeals in this debate contained both technical and what he calls "homocentric" appeals, or appeals to human well-being ("Saving the Great Lakes" 154).

The final chapter of this book, chapter 5, "A Tale of Two *Logoi*: Citizen Science and the Politics of Redevelopment," extends the investigation of the interaction between techno-scientific and nonexpert *logoi* in public policy debate by pursuing the question, To what extent and in what ways might citizen science influence public policy arguments and outcomes? In exploring this question, it examines a citizen-science sound-mapping project in the Pepys Estate neighborhood in the London borough of Lewisham. The mapping project—developed collaboratively by university researchers, nonprofit organizers, and local residents—was designed to measure the noise levels created by a local scrapyard. By weaving together expert techno-scientific *logos* and nonexpert *logos*, resident citizen scientists were able to persuade the borough to put a stop to the scrapyard's noisy activities. Though residents were able to achieve a measure of success in policy debate with these mixed arguments, their influence on the policy process had some unintended consequences. In particular, their scientifically supported conclusions about noise pollution were used by the borough to make the case for building a housing development on the scrapyard site, a policy solution that residents did not fully support. By carefully examining the discourse and policy arguments of both citizen scientists and borough representatives, this chapter extends the conversation on the mixed uses of expert and nonexpert *logoi* by showing that while scientific evidence gathered through citizen science can strengthen lay policy arguments, this same citizen-gathered evidence can be exploited by policymakers to advance their own policy agendas.

Collectively, these chapters recognize citizen science as an important new space for scholarship in rhetoric and rhetoric of science by exploring in detail the ways in which Internet-supported citizen science is transforming interactions and arguments between laypersons, science, scientists, and policymakers. Considering how the Internet has changed the way we argue or communicate with one another is, however, not a completely novel idea. There have been a number of books published on the subject in the last seven years. Barbara Warnick and David Heinemann's *Rhetoric Online: Frontiers in Political Communication* (2012), for example, explores the use of new media in the 2008 election and the role of viral YouTube videos on the legislative debate over Don't Ask Don't Tell. In a similar fashion, Ian Bogost's *Persuasive Games* (2007) investigates the role of the Internet in political messaging by examining how online games have been used to advance partisan perspectives on the American war in Iraq and the famine in Darfur. Though these current volumes contribute to our understanding of the influence of the Internet on communication and argument, neither of them deals with technoscientific issues. The only publication that has investigated these topics is Alan Gross and Jonathan Buehl's recently published edited collection *Science and the Internet*, which offers readers a sustained examination of the influence of the Internet on scientific practice and argument.[5]

A search of the major rhetoric journals also suggests that very few rhetorical scholars have paid attention to the space where the Internet, science, and argument intersect.[6] It reveals that only a single article has been published on the subject.[7] This article is limited, however, to a discussion of the self-representation of laypersons on science blogs. Through the exploration of citizen science in a variety of real-world cases, this book endeavors to extend the study of the Internet and its influence on scientific argument and communication to a variety of complex interactions between laypersons, scientists, and decision makers. In so doing it expands not only on-going conversations in rhetoric and rhetoric of science scholarship but also contributes to the broader inquiry into the possible social consequences of citizen science.

1

Citizen Science at the Roots

To argue that digital technologies are reshaping the interactions between lay-persons, science, scientists, and policymakers through citizen science suggests either that citizen science is a completely novel product of the digital age or that digital-age technologies have introduced some significant differences into the practice of citizen science that can account for the changes in these interactions. This chapter shows that citizen science is not a new phenomenon but rather an enterprise with historical roots. Although the scientific community has only recently adopted the phrase "citizen science,"[1] early instances of the practices that this phrase currently describes can be identified in a diversity of mid-nineteenth- to late-twentieth-century endeavors including meteorological studies and bird counts (Silvertown 467; Bonney et al., "Citizen Science" 978). In addition to showing that citizen science is not new, this chapter also investigates the changes that digital technologies have introduced into it by assessing the similarities and differences between historical and modern citizen science. In particular, it compares the challenges that have faced researchers attempting to work with laypeople and the strategies they have devised to overcome these obstacles. This comparison suggests that though the challenges of doing citizen science are similar across time, the strategies for addressing these challenges have significantly changed with the introduction of digital technologies. I argue that these changes, along with the growing importance of big data, have motivated twenty-first-century scientists to embrace citizen science as a legitimate part of the scientific enterprise. This acceptance, however, has created the potential for new spaces of conflict between laypersons, science, scientists, and policymakers and challenges to traditional conceptualizations of science.

THE SMITHSONIAN WEATHER PROJECT (1848–1870)

Modern citizen science might arguably be traced back to the birth of the *Philosophical Transactions of the Royal Society* in 1665 when its first secretary Henry

Oldenburg circulated a call for correspondents "to impart their knowledge to one another and contribute what they can to the grand design of improving natural knowledge" (Oldenburg 1). In this investigation, however, I trace the predecessors of modern citizen science back to the mid-nineteenth century—a time when science began to institutionalize as governments turned with increasing frequency to science to deal with practical issues of commerce, war, agriculture, and health. It is in this framework of emerging professional scientific identity and institutional involvement in science that the practice of a citizen science involving the participation of nonexpert, noninstitutional actors could emerge as opposed to the more aristocratic gentleman of science of the seventeenth to early nineteenth centuries. It was in this spirit of more democratic participation in science that the Smithsonian Institution was founded with a bequest from James Smithson to support "the increase and diffusion of knowledge." The Smithsonian's first big scientific venture, the study of the weather, epitomized the practice of citizen science because it relied on a mix of government support, scientific expertise, and volunteer labor to achieve its research goals. By examining this early enterprise, it is possible to get a sense of the myriad challenges that accompanied the involvement of laypersons in science as well as the strategies developed to meet those challenges.

Unlike modern citizen science, the Smithsonian's meteorological project was developed to establish the foundations for a scientific field rather than advance the work of an already existent scientific paradigm. In the late 1840s when the Smithsonian project began, meteorology was a fledgling science both in the United States and Europe. There were scattered investigations on different weather phenomena but no agreed-upon paradigm or meteorological institutions to ground the science. In this preparadigmatic period, there were disputes over a variety of issues that involved a range of theories and explanations, few of which could be empirically supported. One of the most famous disputes of the period, for example, was over the nature of storms. This controversy pitted prominent American meteorological researchers William Redfield, James Espy, and Robert Hare, who each held radically different theories of the phenomenon, against one another.[2] In order to make progress on this and other questions involving the weather, scientists realized they needed to collect data on a grand scale. John Herschel, the famous British astronomer and philosopher of science, for example, remarked about meteorology: "[It] can only be effectually improved by the united observations of great numbers widely dispersed . . . [it is] one of the most complicated but important branches of science, . . . [and] at the same time one in which any person who will attend to plain rules, and bestow a necessary degree of attention, may do effectual service" (Herschel 133).

In sympathy with Herschel's call for mass observation to investigate questions about weather, Joseph Henry, the first secretary of the Smithsonian, dedicated a significant portion of the organization's energy and budget to studying storms and other meteorological phenomena. The Smithsonian's efforts from the late 1840s to the 1870s to study the weather illustrate the challenges that faced scientific institutions who wanted to use laypeople to study nature. The first obstacle was funding. Without financial support it would be impossible to gather and process the mountains of data necessary for studying the weather. Fortunately, the Smithsonian was an endowed institution. Henry, however, still had to persuade the Board of Regents that the project was worthy of funding. In 1847 he made his case to obtain funds to "extend meteorological observations, for solving the problem of American storms" (qtd. in Fleming 76). In response, the board allotted him $1,000—nearly 6% of the institution's budget—for "the commencement of meteorological investigations" (qtd. in Fleming 76). Although in the late 1840s $1,000 was a substantial sum for the scientific investigation of the weather, it was not nearly enough money to send trained scientific personnel, even if enough could be found, across the United States to take measurements or pay for their instruments. Instead, it was necessary for the Smithsonian to prevail upon nonexpert American citizens to volunteer their time and even supply their own equipment to aid in the advancement of science. Toward this end, the institution sent a circular out to members of Congress who were asked to distribute them "to such of their constituents who were judged by them to be favorable to the undertaking" (Foreman 68). Out of the 412 persons in 30 states who received the circular, 155 observers volunteered to participate in the project (Foreman 69).

The volunteers who agreed to take part represented a broad spectrum of American society, though there was greater representation of some professional groups than others. Not surprisingly, the initial call for volunteers was embraced most readily by the educated elite, many of whom were already engaged in the scientific pursuits. In a bibliographic analysis of participants, historian James Fleming identified almost half (47%) of the volunteers as being from "scientific, technical, or educational occupations" (Fleming 92). The other half was drawn largely from professions that could accommodate the routine of observation, which required volunteers to record weather conditions three times a day (at 7 A.M., 2 P.M., and 9 P.M.), six days a week. In "Report of the General Assistant with Reference to the Meteorological Correspondence," Edward Foreman, a clerk at the Smithsonian associated with the project, explained, "The observers are generally persons engaged in occupations which admit to some extent of their being present at the place of observation at the required hours of the day all year round. . . . The classes to which the observers belong, are professors in colleges, principles or teachers in academies, farmers, physicians, members of the legal and clerical profession, and a few engaged in mechanical and mercantile pursuits" (Foreman 77–78).

Foreman's description of the volunteers suggests they needed to have regular schedules and reveals that this criterion invited a broad social spectrum of participants. Although many contributors were scientific or educational experts, many others were laypersons without training in meteorological instruments or scientific methods of observation. Over time participation from laypersons increased as the project expanded observation in rural and sparsely populated areas of the western United States and as scientific professionals turned increasingly toward their own research interests and away from what they considered the less profitable Baconian activities of weather observation. Because of these changing dynamics, by the end of the Smithsonian's weather project the percentage of farmers involved in observations had risen from 8% in 1851 to 37% in 1870 while the participation of scientific, technical, and educational observers had fallen from 47% to 16% (Fleming 92).

With such a broad range of expertise represented in their observer pool, another significant challenge facing Smithsonian organizers was to figure out how to get reliable standardized data about the weather from their participants. Henry's initial strategy to encourage standardization was to provide all observers, free of charge, with a set of forms for recording their data. These forms came in three versions, which varied in accordance with the kinds of instruments the observers had. Number one forms were for observers with the most instruments, which often included a wet and dry bulb thermometer, barometer, and rain gauge. Number two forms were for those with a more limited instrument kit, which usually included a thermometer and weathervane. Finally, the number three forms were sent out to those without any instruments at all. Each day observers would enter into their forms data about temperature, barometric pressure, and wind direction as well as information about the type of cloud cover and the levels of rain or snow. At the beginning of every month, they would receive new forms and mail their completed ones back to the Smithsonian.

Though the use of standardized forms disciplined the kind of information the Smithsonian organizers received, it could not account for the accuracy and consistency of the measurements themselves. A particular concern was the standardization, or lack thereof, of measuring instruments and the variety of methods participants used to take and record measurements. In an effort to determine and correct for the lack of standardization in instrumentation, the Smithsonian hired Swiss émigré Arnold Henry Guyot to tour academies in New York to examine the consistency of the instrumentation and the correctness of their measurement practices. In the course of his tour, Guyot was shocked at the dismal quality of the instruments he encountered. He wrote to Joseph Henry in January of 1850, "I have not seen

one station even less an ensemble of stations . . . which operate under circumstances and with instruments much poorer than those I have seen in my
many travels" (qtd. in Fleming 118). As a consequence of his investigation,
Guyot commissioned New York City instrument makers Green and Pike to
design a new standard barometer and purchased thermometers and barome-
ters from the instrument maker to distribute to participants in the Smithsonian meteorological network (Fleming 119–20).

In addition to commissioning and purchasing more accurate instruments,
Guyot also made an effort to improve the standardization of measurement
by writing a handbook for observers that described how to properly set up,
calibrate, and read instruments as well as how to properly record results on
the official meteorological forms. In describing how to read a thermometer in winter, for example, he advises, "The reading should be made at all
times, and especially in winter . . . without opening the window; otherwise,
the temperature of the chamber will inevitably influence the thermometer in
the open air" (Guyot 8). Along with advice about how to accurately take and
record measurements, Guyot also attempted to inspire Smithsonian weather
observers to maintain their discipline in their data collection and daily calculations by prevailing on their sense of civic responsibility: "It is only by making the correction himself [of the average of daily, monthly, and yearly tem-
peratures] that the observer can institute his own comparisons, and really
study the course of meteorological phenomena. His interest will increase
still more with the feeling that he is cooperating in a great work, which concerns at once his whole country and the science of the world, and the success of which depends upon the accuracy, fidelity, and devotion of all who
take part in it" (Guyot 41). As this exhortation illustrates, the correction of
instruments and the preparation of standard charts were not the only strategies relied on for encouraging accurate science. Persuasive, or rhetorical, argument was also employed. By appealing to lay observers' sense of community and duty to science and country, Guyot evokes the spirit of the "citizen
scientist," though not the phrase, to encourage Smithsonian observers to keep
up their data gathering and take care in their efforts.

After a sustained effort to raise money, find participants, ensure accuracy,
and encourage cooperation, the Smithsonian meteorological campaign began
to bear fruit. In 1861 and 1864, the US Patent Office published two volumes
that contained the reduced statistical data gathered by Smithsonian volunteers under the title *Results of Meteorological Observations under the Direction
of the United States Patent Office and the Smithsonian Institution from the Year
1854 to 1859, Inclusive.* Based on the data from this volume, a number of important scientific publications were produced. Most notable, perhaps, were
James Henry Coffin's *Winds of the Northern Hemisphere* (1853) and *Winds of
the Globe* (1875), which offered empirically grounded descriptions of the gen-

eral circulation of the atmosphere across the earth and evidenced-based advice about how to optimize the safety of maritime and naval navigation during hurricanes.[3] Though the work of the Smithsonian weather project did not
ultimately end the storm controversy, it did make American meteorology the
envy of Europeans whose efforts to systematically collect data did not begin
until 1854 and who found themselves far behind the United States in their
capacity to predict and visualize the weather (Anderson 2, 247–49).

By examining the case of the Smithsonian meteorological project, we witness some of the exigencies and obstacles that influenced the organizers of
early citizen science. For Henry and the other organizers of the Smithsonian
project, the primary exigence for engaging laypersons in data gathering for
science was to contribute to the storehouse of scientific knowledge about
the weather and to settle outstanding scientific debates about meteorological phenomena. To accomplish their ends, they depended on the goodwill
and diligence of observers from a broad spectrum of society geographically
stretched across the United States. Because of the uneven spread of material
and epistemic resources across this population and because of their desire
for accurate data, the organizers had to raise money, calibrate instruments,
teach the fundamentals of scientific observation, and exhort their participants
to remain faithful to the practices and the general mission of the project.
Through their efforts, they were able to create a stable network of volunteer
lay observers that provided them with weather data that allowed meteorological theories and laws to be developed and helped meteorology to take
its place as a modern science in the twentieth century. Though creating the
foundations of a new science is a rare exigence for modern citizen science,
the challenges of raising money, ensuring accuracy, and encouraging participation are still very real obstacles. However, as we will see, these obstacles
are addressed in different ways in digital-age citizen science.

THE CHRISTMAS BIRD COUNT (1900–PRESENT)

Although the development of modern meteorology is an example of a spec-
tacularly successful and still on-going citizen-science project,[4] perhaps the
most frequently cited precursor of modern citizen science is the Audubon
Society's Christmas Bird Count. Initiated in 1900 by Frank Chapman, an
ornithologist and curator at the American Museum of Natural History, the
Christmas Bird Count, or Christmas bird census as it was originally called,
offers a glimpse of a different sort of citizen-science endeavor. Unlike the
Smithsonian meteorological project, whose primary goal was to develop a
paradigm for the science of meteorology, the goal of the Christmas Bird Count
was to cultivate interest in and to educate the public about birds and the science of ornithology. With this shift in goal from gathering data for science

about a natural phenomenon to generating interest in one, there is a subsequent change in the challenges for scientists in working with laypeople and in the strategies for overcoming these challenges.

It is not surprising that the goal of the Christmas Bird Count was to educate the public and to generate interest in the study of birds, because its sponsors, Audubon Societies[5] of America, were initially founded as clubs of social activists interested in the preservation of birds. During the late nineteenth century, the founder of the first Audubon Society, George Bird Grinnell, became alarmed after counting the number of stuffed birds and bird feathers he saw adorning the hats of the ladies of New York as he walked from his home to his office. The experience opened his eyes to the need for the protection of birds and inspired him to start the first Audubon Society to advocate for their preservation. Though the society was discontinued because of Grinnell's inability to handle the overwhelming response it received, independent Audubon Societies developed to pursue the issue of bird preservation (Stinson 5). Because of their political focus, these groups were considered by the American Ornithologists' Union (AOU) as social rather than scientific associations. Despite the nonscientific goals of these societies, their leaders were frequently ornithologists, like Frank Chapman, who believed that introduction to the scientific study of ornithology was crucial for getting the public interested in birds and their preservation. In 1899 Chapman became the first editor of *Bird Lore* magazine, the official publication of the Audubon Societies. Under his guidance, the publication developed as an organ for education and outreach about birds and bird issues to the societies' members and the broader reading public.

In early editions of the magazine, we can see evidence of Chapman's dedication to these goals in his writing. In response to a dispute about allowing Audubon members to be initiated into the American Ornithologists' Union, Chapman very clearly articulates the difference between the Audubon Societies and the scientific union. He writes, "We would make no comparison between the Audubon Societies and the [American Ornithologists'] Union. . . . When their relations are properly understood, it will be seen that they stand to each other as preparatory school to college. It is the province of the Audubon Society to arouse interest in the study of birds. . . . It is the province of the A.O.U. to enroll them in its membership after the school day period . . . and sustain their interest through the stimulation which comes from association" (Chapman, "The AOU" 162).

The explicit focus on generating interest in birds and their protection described in this editorial is also evidenced in Chapman's original call for a Christmas bird census. The census itself was designed as a basic ornithological field exercise meant to educate its participants on birds and bird identification. These goals are evidenced in the announcement of the count, which includes a basic set of instructions about what data to collect, an echo of the Smithsonian's weather forms. In describing the categories for data gathering and procedures, Chapman explains, "Reports should be headed by the locality, hour of starting and of returning, character of the weather, direction and force of wind, and the temperature. . . . The birds observed should then be added, following the order in which they are given in the AOU 'Check List,' with, if possible, the exact or approximate number of individuals of each species observed" (Chapman, "Christmas" 192).

Because they require participants in the census to consult a number of instruments (e.g., watch, weathervane, and thermometer) and list species using an official scientific reference guide, Chapman's instructions have a basis in science. Though Chapman is likely drawing on ornithological methods to organize the census's data collection activities, there is evidence that these instructions are directed for the education and enjoyment of the participants rather than the scientific study of bird populations. That the Christmas bird census, unlike the Smithsonian meteorological project, was not designed to answer specific scientific questions is evidenced by the fact that there was no coherent scholarly conversation in the ornithological community about conducting population studies of birds until almost a decade and a half after the census was announced. A search of the content and titles of articles in the American Ornithologists' Union's magazine *The Auk,* using the phrases "bird census" and "bird count," for example, identified no articles dealing with these themes until the announcement of the first national bird census in 1914. In addition, there is no evidence that data from the Christmas bird census plays a role in the scholarly work of ornithologists until 1914,[6] and then even after this date its data does not appear with any regularity in scientific work until the 1930s (Stewart 185).

If the bird hunt was just for fun, then why ask participants to collect data at all? Part of the reason, of course, is education. By learning how to identify and describe birds and their conditions, amateur birders would learn the basics of ornithological identification and be able to make connections between seasonal conditions and avian populations. Similarly, the counting aspect of the exercise would promote awareness in the reader of the number of birds in their vicinity, which was tied closely to the Audubon Societies' political goal of preservation. However, another reason to involve counting was to enhance the entertainment value of the activity. The counting-as-entertainment aspect of the bird count is emphasized at the very beginning of Chapman's 1900 announcement of the Christmas bird census in which he compares the activity with a traditional Christmas "side hunt." A side hunt, as Chapman explains, involves "representatives of the two bands [or sides] . . . [going into] the fields and woods on the cheerful mission of killing practically everything in fur or feathers that crossed their paths" (Chapman, "Christmas" 192). The point of the traditional hunt was to see which of the two sides could score the most kills. As Chapman recalls, the kills of the winners "were often published in our leading sportsman's journals, with perhaps a word of editorial commendation for the winning side" (192). The Christmas bird census was offered as an alternative side hunt wherein the participants would target their quarry with their eyes and ears instead of their guns and record their take by marking numbers on a sheet instead of counting the casualties of the hunt. This allowed the thrill of the hunt and competition to be maintained while at the same time avoiding the decimation of the natural resources the Audubon societies endeavored to protect. Participants in the bird census would also receive their moment in the spotlight for their birding skills, because Chapman published the results of their counts in the January issue of *Bird Lore* each year.

Chapman's bird census turned out to be a great success. The first census in 1900 included a total of twenty-seven participants in twenty-five locations. By 1951 there were 5,151 participants counting in 433 separate locations. In 2012 tens of thousands of bird enthusiasts participated in two thousand locations (Stewart 184; "About the Christmas Bird Count"). As coverage of the count increased, scientists became interested in the data as a source of information about winter bird populations and migration patterns. This rise of interest was accompanied by a concern on the part of the scientists that the methods for gathering the data were not sufficiently robust to ensure the accuracy of the counts. However, unlike the Smithsonian's weather project, scientists could not require participants to gather data in a particular way. Because the bird count was set up by a conservation society for the purpose of educating its constituency and getting them interested in birds and their protection, there was no scientific mandate for the project or power on the part of scientific institutions to control the activities of the participants. The challenge for scientists, therefore, was how to negotiate their goals for getting pristine data against the aims of the Audubon Society and its members for promoting public interest in and education about birds.

Historically, this negotiation has been a delicate one which, more often than not, has required scientists to settle for incremental changes in the way the Christmas Bird Count is conducted. Evidence of the delicacy and limited power of scientists to impress their vision of ideal citizen science on the count emerges in the 1950s, '80s, and 2000s in association with a growing appreciation of the value of its data and an emerging technological capacity to process it. The first scientific call to improve the count appeared in the ornithological journal *The Wilson Bulletin* in 1954. In the article "The Value of Christmas Bird Counts," ornithologist Paul A. Stewart opines, "The Christmas bird count could be a highly effective method of collecting data on early winter bird populations, but the techniques now used are in need of refine-

ment if the data are to have the maximum, or even much, scientific useful-
ness" (193). In order for the count to realize its potential as a scientific tool,
Stewart argued that participants needed to accurately identify the area they
covered in their count and to use the same paths of observation each year. He
also criticized the reliability of observers' identifications and their methods
of canvassing an area for birds.

Stewart's suggestions for changing the count were met with eloquent indignation by a longtime bird-count participant Joseph Hickey, a professor
at the Department of Forestry and Wildlife Management at the University
of Wisconsin. He responded to Stewart's comments in a letter to the editor
of *The Wilson Bulletin*: "May I submit that Christmas bird counting is based
upon a largely emotional component that still has a reputable place in our
largely materialistic society today? . . . However much we may wish for other
improvements in the Christmas count, let us recognize that many of these
suggestions impose disciplines that most laymen will simply not accept.
The Christmas bird count is to them essentially a recreational activity in
which distinct elements of competition, surprise, rarities and the big list are
bright and personally fulfilling" (Hickey 144). Hickey's argument—that making the bird count scientific would rob it of its fun and excitement—illustrates clearly the outsider status of scientists in the bird count and the obstacles they faced in making it scientific. In particular, Hickey suggests here
that placing too much discipline on the count might discourage people from
participating, which would not only destroy an important social mechanism
for educating laypeople about birds, but also diminish the utility of the count
as a source for data.

Similar precarious negotiations took place in the 1980s and again in the
first decade of the twenty-first century. In the 1980s statistical researchers
suggested that they might work with the Audubon Society to promote an
"ideal model" for data collection and processing. To persuade participants
to cooperate with scientific schemes for better data-gathering practices, researchers prevailed, just as Guyot did, on their sense of scientific and civic
pride: "An ideal model count obviously demands greater effort and care by
every participant especially by compilers. . . . Various direct incentives will
be offered [for] Ideal Model counts: forgiveness of 50% of participant fees,
[and] the honorary designation of 'Elite Counts.' The incentives of pride in
leadership, of pioneering into new frontiers, of acquiring reputations of superiority are powerful motivating forces" (Arbib 148).

Despite the civic incentives of being considered a leader or a pioneer in science, the ideal model never caught on with the participants of the Christmas
Bird Count. By the first decade of the twenty-first century, scientists' hopes
for a more disciplined data collecting regime were greatly diminished. They
resigned themselves to the fact that the burden for fixing errors in the bird

count would fall mainly on them. Toward this end, scientific institutions interested in the data, like the Cornell Lab of Ornithology, began to invest in
statistical resources for assessing count biases and developing mathematical
corrections that would eliminate them. As for improving the contribution
of lay volunteers, Dunn and coauthors of "Enhancing the Christmas Bird
Count" (2005) suggested that the Audubon Society leadership might 1) "require participants to separately record birds seen during feeder watching from
indoors . . . and other specialized counts" and 2) "establish standardized count
areas or routes within each [watch] circle" (342). Even these few suggestions
were presented as a scientific wish list rather than as edicts, a sign that researchers accepted that forcing the activity into a scientific mold might jeop-
ardize its educational and social value: "Although it is theoretically possible
to revise the CBC [Christmas Bird Count] protocol to address all the challenges related to site selection and effort variation, the panel concluded . . .
that changing survey design and data-collection methods to mold the CBC
into a rigorous population-monitoring program would significantly reduce
its value. . . . Important nonscientific benefits of the CBC would be lost, including social and educational benefits to participants and the community
at large and value as an entry point for citizen science" (Dunn et al. 342).

The brief history of the Christmas Bird Count illustrates that some traditional citizen-science projects and their digital-age descendants were not developed to achieve the goals of the technical sphere. Instead, it shows how
public sphere needs to persuade people to value and protect natural resources
can also encourage lay participation in gathering information about nature.
Despite its humble beginnings, the Christmas Bird Count has become one
of the most successful efforts at public outreach in the history of the conservation movement. Because of its popularity and its semiscientific format
of collecting information on birds and the conditions in which they appear,
the data from the count has attracted scientific attention and interest. With
this interest, the shortcomings of the count's methods have become a subject of scientific inquiry. This inquiry illuminates another challenge shared
by historical and modern citizen science: the tension between the needs and
values of laypersons gathering data and the needs and values of scientists
who compile and process it.

CITIZEN SCIENCE IN THE DIGITAL AGE

The historical examples discussed in the previous sections suggest that there
are similarities between what we now call "citizen science" and early efforts
to enlist laypeople to gather information about nature prior to the digital
age. These similarities are recognized by twenty-first-century scientists who
frequently place their own citizen-science research in the context of these

earlier projects. In a 2009 roundtable discussion on citizen science, for example, researchers at the Cornell Lab of Ornithology write, "Public participation in scientific research is not new. Lighthouse keepers began collecting data about bird strikes as long ago as 1880; the National Weather Service Cooperative Observer Program began in 1890; and the National Audubon Society started its annual Christmas Bird Count in 1900" (Bonney et al., "Citizen Science" 978).

Despite acknowledging affinities between modern citizen science and earlier data-gathering activities, scientists in the twenty-first century have argued that there are significant differences between the two enterprises. A qualitative examination of recent scientific articles about citizen science reveals a number of ways in which scientists distinguish modern citizen science from its historical counterpart. These differences include changes in the scale of the research, the kinds of laypersons involved, the strategies for retaining volunteers, the methods of assuring data quality, and the sources of project funding. In a 2007 discussion of the topic, for example, Caren Cooper and her colleagues at the Cornell Lab of Ornithology comment on modern citizen science's potential to change the scale on which environmental problems might be dealt with. They write, "Combining the power of the Internet with a populace of trained citizen scientists can provide unprecedented opportunity to mobilize a community to address new environmental problems, almost like having the environmental equivalent of a "fire brigade" ready to act as the need arises" (Cooper et al. 8).

While some researchers have distinguished modern citizen science from its historical counterpart by the number of participants that can be mobilized, others have identified its capacity to expand the kinds of laypersons who can be involved in scientific activities as a point of difference. Jonathan Silvertown, a British professor of the life sciences, argues in the journal *Trends in Ecology and Evolution*, for example, "The characteristic that clearly differentiates modern citizen science from its historical form is that it is now an activity that is potentially available to all, not just a privileged few" (Silvertown 467). This broadening of the participant pool, he argues, is related to "the existence of easily available technical tools for disseminating information about projects and gathering data from the public" (467). Whereas the Smithsonian meteorological project relied on tracts circulated by members of Congress amongst those judged "to be favorable to the undertaking," advertising for modern citizen-science projects takes place online. As a consequence, digital-age scientists have the capacity to reach out to thousands and even tens of thousands rather than hundreds of potential volunteers. Though modern calls for participation typically find their way to segments of the public who are already interested in wildlife, conservation, or the sciences, the presence on the World Wide Web of citizen-science projects allows for the possibility that at least some users outside these interest groups might become involved.

In addition to potentially reaching a broader and more diverse range of people, digital-age citizen-science projects have access to new methods for keeping these participants engaged and ensuring their work meets scientific standards for quality. Though the organizers of the Smithsonian weather project gave volunteers free copies of the compiled weather statistics as a token of appreciation for their efforts, there was a substantial lag-time between their labor and reward. Observers who contributed data to the project in 1854 didn't receive the *Results of Meteorological Observations* until 1861, seven years after they had made their contributions. In the age of the Internet, feedback can come in days, minutes, or even seconds. On the Cornell Ornithology Lab's eBird website, for example, citizen-science bird watchers are given instant access to their data, the data collected by other participants, and the tools to compare them ("About eBird"). These website features have been so successful at increasing participation that researchers reported that "immediately after these features were implemented, the number of individuals submitting data nearly tripled" (Bonney et al., "Citizen Science" 981).

While some digital tools have been developed to instantly reward volunteers and keep them participating in citizen-science projects, others have been created to improve the quality of the data gathered by laypersons. In predigital-age citizen-science projects, scientific organizers relied heavily on paper forms to discipline observations. New online forms, however, not only ensure consistency in data reporting but also automatically prescreen data for anomalies and filter them out before they get into databases (Bonney et al., "Citizen Science" 980). They allow researchers working with data from bird observations, for example, to review records and decide whether a volunteer bird watcher has misidentified a species or whether they have indeed sited a bird that was anomalous for their area. In addition to flagging possible errors, scientists also have the capacity to statistically manage the data to compensate for the problems of quality. With the help of modern computing power, researchers can, for example, combine and compare data from scientifically controlled field studies with less formal observations by citizen scientists to create more accurate models of phenomena like avian disease transmission and nesting times (Bonney et al., "Citizen Science" 981). This strategy for dealing with unruly data would have seemed like an impossible dream for James Henry Coffin and his fifteen statistical clerks in the US Patent Office who worked almost twenty-nine thousand hours to calculate, by hand, the basic statistics for weather observations from 1857–1860 (Fleming 126).

Finally, modern citizen science can be done more cheaply and can tap a broader range of funding sources than its historical counterpart. As in the past, volunteers provide scientists with a free labor force for conducting their investigations. However, modern citizen science has drastically reduced the cost of working with volunteers. Whereas the Smithsonian had to pay to print reporting forms and depend on American taxpayers to defray the costs of mailing these forms, the Internet has made these costs virtually negligible. Further, because of the connectivity of computer hardware, researchers at institutions like SETI have devised ways of borrowing users' computers to boost their computing power rather than having to shoulder the cost of additional servers or supercomputers. These financial benefits of citizen science are recognized by Silvertown who includes them as factors that explain the ascendancy of citizen science in the first decade of the twenty-first century. He writes, "A second factor driving the growth of citizen science is the increasing realization among professional scientists that the public represent a free source of labor, skills, computational power and even finance" (Silvertown 467). Silvertown also links the economic benefits of Internet-enabled citizen science to its capacity to promote social outreach. This capacity, he explains, has made it attractive to scientists competing for government funding, most of which includes outreach as a criterion for awarding grants. Silvertown explains, "Citizen science is likely to benefit from the condition that research funders such as the National Science Foundation in the USA and the Natural Environmental Research Council in the UK now impose on every grant holder to undertake project-related science outreach. . . . If we want to continue to spend taxpayers' money, it is in scientists' own interest to make sure that the public appreciates the value of what they are paying for" (469). In addition to making citizen science more attractive to government grant givers, the Internet has, perhaps more importantly, made the diversification of sources for funding citizen-science projects possible. Crowdfunding sites like Kickstarter, for example, allow scientists and even laypeople interested in doing citizen science to find financing for their research outside of the traditional institutional sources like governments and foundations.

Although scientists have identified a variety of ways that modern citizen science differs from its historical counterpart, their assessments all point to a common source for these differences: the development and expansion of digital technologies, particularly the Internet. In the last decade and a half, digital technologies have made working with laypeople more practicable and attractive to scientists. As these conditions of doing science have changed, so too has the status of citizen science. Once a relatively uncommon enterprise without a formal title, citizen science has now become a regularly practiced and officially recognized category of sanctioned scientific activity. A search of scientific journal databases,[7] for example, shows that citizen science appears as a topic of regular discussion in the scientific literature starting around

2006.[8] This newfound recognition is not the result of the emergence of a new kind of activity, as the historical examples suggest, but rather a consequence of the increase in the payoffs and a decrease in the obstacles of traditional lay data collection made possible by the development of digital technologies.

As digital technologies have accelerated the integration of citizen science into mainstream science and expanded the number and kinds of laypersons that can participate, it is important to consider whether this integration and expansion has consequences for the relationships between laypersons, science, scientists, and policymakers, and for traditional perspectives on what constitutes science. In the cases examined in this chapter, we find evidence that there may be significant consequences. In the history of the Christmas Bird Count, for example, we see how the introduction of modern computing made the count's data more valuable to scientists. As the value of the data increased, so did scientists' efforts to exert control over it. This created friction between the count's lay participants' interest in public outreach, education, and enjoyment and scientists' goal of getting clean data. In pursuit of better data, scientists needed to address this friction by entering into negotiations with the Audubon Society and its members taking them beyond the traditional boundaries of scientific practice.

Though we see hints in the history of citizen science that digital technologies are influencing the relationships between laypeople, science, and scientists and changing the practices of science, little effort has been made to study these influences or changes.[9] The case studies in the chapters that follow provide a detailed exploration of the language and argument generated by digital-age citizen-science projects. These explorations reveal that modern computing and the Internet have opened up the practice of science and the possibilities for laypersons to participate in it in unprecedented ways. They have, for example, allowed for grassroots citizen-science projects—generated from and centered on the interests of laypeople—to emerge granting laypeople unprecedented access to arguments in the public and technical spheres. In turn, however, this expanding access to science has created a number of interesting and unprecedented challenges. For laypeople, digital-age citizen science has generated new situations in which they must struggle to achieve credibility and fight to ensure that their data are not used to promote policy agendas incommensurate with their interests. For scientists, citizen science has drawn them into political and social controversies they may have previously avoided. For policymakers, citizen science allows representations of technological risk to emerge that challenge official accounts of these phenomena. Finally, citizen science disturbs traditional expectations about science, blurring the lines between scientists and nonscientists, implicating the former in social and political agendas and permitting the latter to gather and share data in unprecedented ways and on previously unimagined scales.

Though laypersons, scientists, and policymakers have begun to contemplate the challenges of incorporating citizen science into scientific practice, there are a variety of issues with which they must still grapple. By examining the discourse and argument generated by digital-age citizen science in a variety of circumstances, the following chapters explore both the promise and challenge of digital-age citizen science and its potential to influence science.

2

Reimaging Risk

Citizen Science and the Development of
Citizen-Centered Radiation Risk Representations

In *Risk Society* (1992) Ulrich Beck argues that the central social concern in twentieth-century Western nations is the rise of risk brought about by their regimes of scientific progress and technicization. This insight seems to be reaffirmed almost daily as events like the oil pipeline blowout at the Deep Water Horizon and the meltdown of reactors at Fukushima attract media attention and generate public debate and discourse about the risks of modern techno-industrial society. Rhetorical scholars of science and technology have been quick to pick up on and investigate the growing importance of risk and its consequences for argument and communication. They have written on a range of risk topics including nuclear accidents (Farrell and Goodnight 1998), mine safety (Sauer 2003), and bioweapons (Keränen, "Viral Apocalypse" 2011). They have also been concerned with the problems of modern risk assessment and deliberation. Some rhetorical scholars have examined the way that science-driven policy toward risk obstructs the democratic process by putting it out of touch with the needs of the lifeworld[1] (Katz and Miller 1996; Grabill and Simmons 1998; Simmons 2007). Others have explored the inequity of epistemological authority between technical and social knowledge and have considered how to reclaim influence for public knowledge and reasoning (Fisher 1987; Fischer 2000; Kinsella 2004). Collectively these efforts involve a reimagining of how the relationship between publics and institutions might be reconstituted to support public engagement and reassert democratic control over risk definition and management.

Despite the significant attention to written and spoken risk discourse and argument, virtually no consideration has been paid to visual representations and their role in characterizing risk.[2] This chapter engages with this research gap by exploring maps visualizing the radiation risk from nuclear plant accidents at Three Mile Island, Chernobyl, and Fukushima in both print and online media. In particular it will pursue the questions: Are public representations of risk created by grassroots citizen-science groups different from risk representations in the mainstream media? If so, what might these differences reveal about a grassroots citizen-science perspective on risk communication? Because the mainstream media was the only public source of visual representations of risk following these nuclear accidents and because the visuals were in many cases created with the support of expert sources, they represent an appropriate point of contrast for thinking about the differences between institutional and noninstitutional visual risk representation. To assess the strategies in the print media for representing radiation risk and the differences between institutional and grassroots citizen-science risk representations, this chapter will analyze the visual/verbal features of radiation maps in pre-Internet print stories in the *New York Times* and *Washington Post* on Three Mile Island and Chernobyl. These visualizations will be compared with online maps of the radiation risk from Fukushima created by the *New York Times* and the citizen-science group Safecast to gain insight into whether and in what ways the Internet and grassroots citizen-science groups with access to the Internet might be influencing risk communication about radiation and what the consequences of this transformation might be.

METHOD

Describing and comparing visual representations of risk across nuclear accidents requires analytical categories for visual assessment and methods that allow similarities and variations to be accounted for. To maintain comparability and a manageable sample size, the corpus of print visuals for this investigation has been limited to the *New York Times* and the *Washington Post*. The online visual corpus has been limited to the websites NYTimes.com and Safecast.org. These corpora offer access to important slices of mainstream print and online visual representations of the three nuclear accidents. In conducting this investigation, every issue for each of the print publications was searched for risk visuals for a one-month period following each nuclear accident. Once located, all of the visuals in the sample were assessed using a standard format. Basic background information about the visual was recorded including the date, source, author, and location in the news publication. Then, a detailed qualitative evaluation of the visual was made, which included four categories of assessment: the *format* of the visual presentation, the *information* provided about radiation risk, the *relationship between the visual and the text* of the news story (or stories) with which it was associated, and the *context* in which it occurred. In the assessment of the online risk visuals on NYTimes.com and Safecast.org, these strategies of analysis were also applied with the exception of the analysis of the relationship between the text and the story. The map on NYTimes.com was linked as a resource to a number of stories rather than attached to a single one, making it difficult to identify any specific connection between the reporting and the interactive visual. Safecast.org was not a news site and, therefore, had no news text to analyze. However, maps and blogs posted on the website between March of 2011 and 2012 were examined.

In assessing the *format* of the visual presentation in all of the sources, the range of strategies for representing risk was identified and inventoried. These included features like maps and map insets, visualizations of population centers, visualizations of radiation and its magnitude, and the incorporation of words and numbers in visuals. Once the visual formats were identified, they were assessed for the *type of information* about risk they communicated. This assessment was made on the basis of a basic set of questions considered vital by media experts on risk reporting: What is the risk? What is its magnitude? What is its location and geographic extent? Who is affected by it? What are the consequences for those who are affected? Who/what is responsible for it? (Ropeik 2011; Kitzinger 2009). In cases where risk visuals were complements of news stories or blogs, the *relationships between the visual risk representations and textual content* were also examined. The relationship between words and visuals and the kinds of epistemological and rhetorical contributions each made to discourse and argument has been the subject of research by a number of rhetoric and communication studies scholars (Kress and Van Leeuwen 2006; Hagan 2007; Gross 2009). This study focuses specifically on the extent to which the basic questions about risk were supplied by the text of a news story, its associated visual, or both. Then, conclusions are drawn about the role of visual and textual elements or their collaboration in communication about risk. Finally, this assessment considers the *context* in which the risk visualizations occurred. As Birdsell and Groarke (2007) point out, visual representations need to be interpreted in a manner that fits the context in which they are situated (104). Because the contextual factors influencing risk visualization of nuclear accidents are myriad and not all relevant, this investigation focuses specifically on assessing the influence of existing visual conventions on choices of risk representation, the immediate historical-political context in which the risk visuals emerge, and the technological-material factors which might affect them.

BULL'S-EYES AND CLOUDS: VISUALIZING RADIATION RISK BEFORE THE INTERNET

Before the Internet, the publicly circulated representations of radiation risk from nuclear accidents were dominated by, if not the exclusive domain of, institutional communicators in the mass media, government, and scientific communities. This domination was the consequence of the challenges associated with creating these representations, which required their developers to have the capacity to measure, visualize, and broadcast information about radiation risk. At the Three Mile Island and Chernobyl accidents, only governments, industry, and international organizations such as the IAEA had the ability to measure and track the spread of radiation. In addition, only the mainstream media in cooperation with the government and industry created broadly publicized risk visualizations of the accident. Because the production of risk visuals for the first two major nuclear accidents was monopolized by these institutional actors, visual representations of radiation risk from this period are an ideal resource for identifying and understanding the conventions of institutional risk representation. To illuminate these conventions, the sections that follow explore the risk graphics in the *New York Times* and the *Washington Post* reporting on Three Mile Island and Chernobyl. By describing the visual conventions, exploring their context, and assessing their use as strategies for communication and argument, these sections provide a touchstone for the comparison in the second part of the chapter between mainstream media and grassroots citizen-science risk visualizations.

THREE MILE ISLAND AND THE BULL'S-EYE OVERLAY

At 4:00 A.M. on Wednesday, March 28, 1979, the main cooling pump of reactor number two at the Three Mile Island nuclear plant shut down and the auxiliary pump could not be brought on line to cool the reactor. The plant's engineers shut the reactor down; however, pressure built up in its core as uranium rods continued to fission without being cooled. Steam building in the reactor core was released to maintain pressure, but a stuck-open valve allowed cooling water to flow out of the reactor. A cascade of events followed, including the melting of more than half of the reactor core; the buildup of a hydrogen gas bubble in the reactor; and the release of radioactive water, vapor, and particulates into the area surrounding the plant. As the crisis grew, the news media descended on southeastern Pennsylvania to cover the event. Three Mile Island represents a watershed moment in radiation risk communication, because it was the first primetime nuclear disaster for which writers and visual designers in the mainstream media were tasked with representing radioactive risk. Though there were two other accidents reported on in the 1960s,[3] these did not generate the media attention or risk visuals that the Three Mile Island accident did (Gamson and Modiglioni 14). Because of their uniqueness, the verbal/visual risk representations in the media coverage of Three Mile Island offer a starting point from which to assess the developments of mainstream media conventions for visualizing radiation risk from nuclear accidents.

Though the visualizations created to describe the accident at Three Mile Island were the first of their kind, it is important to note that the accident did not immediately or directly spawn new conventions for representing radiation risk. Evidence that conventional representations were not immediately created can be found in the first two days of reporting (March 29 and 30, 1979). During this period, the *New York Times* and the *Washington Post* both carried maps of the area at risk from the accident. However, only the *Washington Post* visualized radiation on its map using a shaded square to identify the supposed area affected by radioactive emissions (Furno). It wasn't until March 31 when the *Washington Post* and the *New York Times* simultaneously introduced maps with a bull's-eye overlay—a set of concentric rings radiating out from a central point—that a standard for representing radiation risk for the accident was adopted. After March 31 this visual appeared on every radiation map in the sample.

Rhetorical scholars have argued that visual strategies seldom emerge without precedent. Instead, existing conventions for representation in one area are often borrowed or tweaked to create new visual regimes (Kostelnick and Hasset 7). In the case of the bull's-eye overlay, its adoption as a standard for describing nuclear accidents seems to have been encouraged by earlier civil defense evacuation and risk assessment maps, which used the bull's-eye graphic to represent the area of risk created by a hypothetical atomic bomb attack and its consequent fallout. Visualizations of this kind appeared as early as 1952 in the *Greater Boston Civil Defense Manual* and continued to be part of the format of local[4] as well as national[5] civil defense pamphlets and booklets distributed to the public into the 1960s. Though by the 1970s civil defense against nuclear attack was no longer a priority of the government[6] and the circulation of these documents ebbed, their use for a decade in public education had likely made the bull's-eye overlay a familiar visual representation of radiological risk.

In addition, the bull's-eye overlay was also used in mainstream reporting during the 1950s and '60s to describe the threat of a potential nuclear strike. In 1955, for example, the concern over the consequences of nuclear warfare reached a fever pitch with the release of estimates of the area of destruction created by the hydrogen bomb in the 1954 Bikini Atoll tests. These tests revealed that the hydrogen bomb was more powerful than anticipated and the spread of fallout greater than expected. In the first few months of 1955 following the test, stories such as "U.S. H-bomb Test Put Lethal Zone at 7,000 sq. Miles: Area Nearly the Size of Jersey Covered by Atom Fallout . . . Civilian Peril Stressed" and "City Evacuation Plan: 3 Governors and Mayor Weigh Plans to Meet H-Bomb Attack" were front page news in the *New York Times* (Blair; Porter, "City Evacuation"). In the same period, the Associated Press (1955) produced a map, "Radiation Effects," with a bull's-eye overlay to describe the "range of possible death if an H-bomb should hit squarely on Cincinnati." In the visual a bull's-eye overlay with three concentric circles marks off three zones of radiation risk from fallout in the aftermath of a nuclear attack. In the inner 140-mile radius circle, all persons downwind from the bomb blast could expect to receive a fatal dose of radiation. In the two rings marked off between 160 and 190 miles, five to ten persons out of one hundred exposed to radiation might be expected to die. Because of the bull's-eye overlay's pervasiveness in civil defense materials and mainstream media reporting and its connection with the risks of radiation, it is not surprising that it would be adopted to represent the radiation risk from the nuclear accident at Three Mile Island.

Though the bull's-eye overlay was uniquely suited for representing the risks associated with a hypothetical nuclear strike, there were some consequences for choosing this method to represent a real radiological disaster. In civil defense materials, for example, the bull's-eye graphic was useful because it could simultaneously represent a number of different dimensions of the risk situation including the actual risk, the area affected by the risk, and/or the area where risk intervention had to take place. Its capacity to embody multiple representations of risk was useful in civil defense documents dedicated to educating the public about the risks of a nuclear strike as well as informing them who would be affected and what they should do in the case of an attack. Under real emergency conditions where simple and direct communications are essential, however, this capacity to be read multiple ways could be detrimental.

An assessment of the first bull's-eye overlay maps in the *New York Times* and *Washington Post* on March 31 illustrates the problem of multiple associated meanings for visual interpretation (Cook, "Area Surrounding"; Lyons). In both the *Washington Post* and the *New York Times*, maps with bull's-eye overlays were juxtaposed with text discussing radiation levels at the plant and text about Pennsylvania Governor Richard Thornburgh's evacuation plans. In the *New York Times*, for example, the first fallout map is next to the headline "U.S. Aides See a Risk of Meltdown at Pennsylvania Nuclear Plant; More Radioactive Gas is Released," which focuses on the extent of the radiation release. Its caption, however, describes Governor Thornburgh's advisement of evacuation (see fig. 1).

The juxtaposition of the map with text describing these two different subjects raises questions about which of the topics the map is representing: the area of radioactive fallout or the area of evacuation ahead of the foreseeable threat from radioactive fallout. This ambiguity is further encouraged by the fact that neither the *Times*'s nor the *Post*'s maps are labeled to indicate whether the area within the bull's-eye represents the area of evacuation or the area affected by radiation released from the plant. The *Post*'s map is titled generically "Area Surrounding Three Mile Island Nuclear Plant" while the *Times*'s map is not labeled at all. Though there is no direct reader-response evidence that the multiple possible interpretations of the maps created confusion or

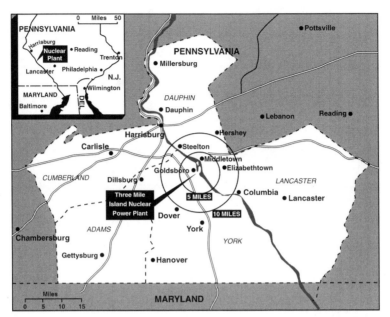

Figure 1. Bull's-eye overlay of the Three Mile Island accident. (Reproduction of a map from the *New York Times*; see *Bull's Eye Overlay*.)

panic, the Carter administration's *Kemeny Report*, which examined the role of the media in the accident, suggests that it might have. The creators of the report comment, "A few newspapers . . . did present a more frightening and misleading impression of the accident. This impression was created through headlines and graphics, and in the selection of material to print" (The President's Commission 58).

In addition to being ambiguous because of its association with multiple meanings, the bull's-eye overlay was also a strategy that offered a generalized rather than a specific representation of risk. The concentric circles that are the hallmark of the visualization represent the possible zones of risk rather than specific details about actual risk location and intensity. In hypothetical nuclear-attack scenarios described in civil defense materials, this representational strategy was necessary since no event had taken place. In fact, imagining the broadest range of potential risk was useful for citizens and local governments who needed to plan for a range of disaster scenarios. While the generalizing quality of the bull's-eye overlay's concentric rings worked well in civil defense manuals, in cases with real risks with specific locations, it could offer a false sense of either risk or security. Readers of the *Times* and *Post*, for example, might assume after engaging with their visuals that all the people living in an area circumscribed by a ring of a bull's-eye overlay would be exposed to the same amount of radiation. In reality, however, radioactive

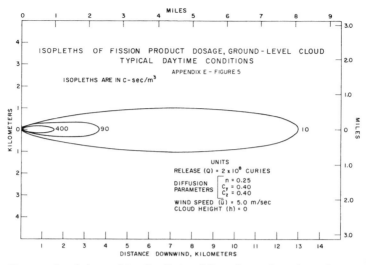

Figure 2. Isopleth graphic of radiation fallout from a hypothetical nuclear accident. (See Atomic Energy Commission 61.)

gases and particles escaping a nuclear plant would follow the path of prevailing winds distributing the risk of radioactivity asymmetrically across an area (ApSimon and Wilson 43). Further, the concentric circle design encourages readers to assume that people living in different rings are exposed to doses of radiation that vary in some regular degree of magnitude from those in the rings further away from or closer to the center of the bull's-eye. Under actual conditions, however, exposure rates are never regular. They tend to be extremely high in areas close to the accident and fall off as distance increases (Von Hippel and Cochran 18). This misconception might be easily cleared up by including radiation measurements in the visual; however, neither the *Post*'s nor the *Times*'s maps included these. In fact, only one mainstream media source, *Newsweek*, provided quantitative values for radiation in their visualizations of the Three Mile Island accident (Matthews et al.).

The obvious communicative drawbacks of the bull's-eye overlay raise questions about why the mainstream media would have adopted this strategy for representing risk or at least not tried to supplement it with numerical data and textual information. More accurate ways of representing radiation risk existed long before the accident at Three Mile Island. As early as 1957, for example, the Atomic Energy Commission (AEC) was employing basic isopleth illustrations—nested lines with assigned values—to describe the direction and concentration of radionuclide releases from hypothetical nuclear plant accidents. Figure 2, for example, shows the hypothesized diffusion of radioactive material under daytime conditions from a ground level cloud pushed by a five-meters-per-second wind (AEC 61).

The figure looks like a bull's-eye overlay whose rings have been pinched and stretched to make concentric conjoined ovals. Though geometrically this isopleth design and the bull's-eye overlay appear similar, their slight differences have important implications for the kinds of information they convey. The pinch and stretch of the isopleth allows it to represent more precisely the general physical shape of the radioactive cloud emitted from the plant whereas the perfect circles in the bull's-eye overlay are visually ambivalent about the direction of the spread of radiation. Further, isopleth illustrations were not limited strictly to internal government publications, which meant they were familiar representations in the broad public discourse about radiation risk. For example, many of the popular civil defense educational materials, such as the film *Radiological Defense* (1961) and the Defense Department's handbook on nuclear attack, *Fallout Protection* (1961), used isopleths to describe the complex shape of fallout.

The existence of more sophisticated ways of visualizing radiation and their presence in popular representations of radioactive fallout deepens the mystery of why the press did not rely on these kinds of visualizations to describe the accident at Three Mile Island. The absence of these types of representations in the daily newspapers likely had more to do with the unavailability of data in the immediate aftermath of the accident than with the media's negligence. One of the criticisms of the Nuclear Regulatory Commission and the utility Metropolitan Edison following the accident was that they had failed to gather sufficient data about radiation levels near the plant and also withheld information about these levels from the public. The scarcity of data is evidenced in news reports on the day following the release of the first bull's-eye maps. On April 1 Walter Pincus of the *Washington Post* reports:

> The Nuclear Regulatory Commission yesterday began to blanket the countryside with devices to monitor and record accumulated radiation.
>
> Until yesterday monitoring outside the plant had been primarily spot checks which determined radiation levels at a specific time and place.
>
> Without knowing accumulative dosages, no determination can be made on the levels of exposure of persons in the vicinity of the power plant. (Pincus A1)

Without accurate comprehensive measures of the radiation levels by a distributed sensor network, it would have been beyond the media's ability to calculate the magnitude of the radiation released from the plant and difficult, though not impossible, to estimate the direction of those releases. As a consequence, the bull's-eye overlay with its capacity for generalization likely offered the most suitable method to describe the risk of radiation emanating from the plant.

An examination of the available visual conventions and the context in which representations of radiation risk were developed at Three Mile Island suggests that the mainstream media, though incomplete and inaccurate in their representations of risk, nonetheless, seemed to have made a reasonable decision when they adopted the bull's-eye overlay. Though they might have done more to address the ambiguity generated by the visual strategy, its familiarity and capacity to communicate generally about risk when information was scarce made it an appropriate and useful visual strategy for describing the accident. As we will see, however, the mainstream media's continued use of the bull's-eye overlay, despite its shortcomings, will invite criticism from the citizen-science group Safecast.

Chernobyl and the Cloud Visual

Whereas the bull's-eye overlay was the dominant convention for risk representation in the mainstream media coverage of Three Mile Island, reporting on the Chernobyl accident included a more diverse visual repertoire. Though the bull's-eye overlay still played a central role in risk visualization, it was joined by the cloud visual, which provided more detailed information about the movement of radiation and its location. This change in the repertoire, while addressing the informational shortcomings of the bull's-eye overlay, was also significant, because it gave the mainstream media a tool for responding to a major rhetorical opportunity occasioned by the accident: the chance to reinforce American perspectives about the Soviet Union. In order to understand the use and affordances of the cloud visual as an ideological tool, this section examines media narratives about the Soviet Union generated from the accident and the way in which cloud visualizations were used to support and advance these narratives.

The accident at Chernobyl occurred on April 26, 1986, at 1:23 A.M. Moscow Standard Time (MST) during a test of the plant's electrical backup systems. The test, conducted under suboptimal conditions, resulted in a spike of radioactivity and heat that blew apart the reactor core. Subsequent explosions and the extreme heat created by radioactive material in a state of uncontrolled fission set fire to exposed graphite moderator blocks in the core sending black radioactive smoke billowing from the plant. The nearby town of Pripyat was evacuated the next day, but there were no reports of the event in the mainstream Russian media until the evening of April 28 (Luke). The first stories in the American media appeared on April 29, three days following the accident. On the same day, the first visualizations of radiation risk appeared (Gwertzman A1).

Communication scholars writing at the time of the accident suggest that the Western media, particularly in the United States, endeavored to make the disaster into a morality tale about US and Soviet cultures (Dorman and

Hirsch 56). One of the moral narratives that dominated the American media was that the accident revealed what the United States and other Western European Countries claimed to have long known about the Soviet Union: that it was a secretive and untrustworthy nation whose leaders were willing to sacrifice their own citizens and put others in danger for the advancement of their social-political agenda. These narratives of *secrecy* and *indifference* were supported by the fact that the Soviets did not release news about the accident to the West until two days after it occurred. They were further strengthened by the circumstance that this release of information was made only after workers at a nuclear facility in Forsmark, Sweden, had detected high levels of radiation and traced it back to the Ukraine. This prompted Swedish government officials to contact Moscow and request information about the event. Only after this request did Radio Moscow announce the accident on the evening of April 28. In the days that followed, reports coming out of the Russian media were terse and defensive. It was not until about a week after the accident that details about the meltdown flowed more freely from the Russian press (Amerisov 38).

Critiques of Russian secrecy appeared immediately in US mainstream reporting on the incident in the *New York Times* and *Washington Post* sampled in this analysis. In the earliest days, the reporting on secrecy was focused on fact. On the first day of coverage, April 29, for example, both the *Times* and the *Post* used reported statements of the Swedish Energy Minister Birgitta Dahl to underscore the fact that the Soviets had withheld information about the accident. In the *Post*, for example, Dahl was reported to have said that "it was 'unacceptable' that Swedish authorities and others outside the Soviet Union had been given no notification" (Bohlen A1). Similarly, the *Times* reported that Dahl "said that whoever was responsible for the spread of radioactive material was not observing international agreements requiring warnings and exchanges of information about accidents" (Schmemann, "Soviet Announces" A1).

In the days that followed, the fact that the Soviets had been secretive about the accident was reiterated in the Western press; however, reporting also began to delve into the causes and consequences of the Soviet cover-up. The *New York Times* article "The Soviet Secrecy," for example, suggested that the Soviet government's decision not to release information was a consequence of its need to maintain an image of control and protect itself from attacks by a hostile Western press: "To the world outside, almost as striking as the nuclear accident that sent radioactive debris over hundreds of miles, was the extraordinary Soviet effort to restrict information about it. It was a reflexive retreat into secrecy that again seemed to show the Kremlin loath to concede any failing before its people and a hostile world" (Schmemann, "Soviet Secrecy" A1). While the *Times* focused on the causes of the cover-up, the *Wash-*

ington Post reported on its consequences. In particular, it examined the consequences of the accident for the Soviet Union's new policy of glasnost, or openness, and the impact their secrecy about the accident would have on its relationship with its Western European neighbors: "The environmental disaster at the Chernobyl nuclear power plant is also rapidly turning into a public relations disaster for the new Soviet leader, Mikhail Gorbachev, who has been trying to impress European opinion with his pragmatism and openness. . . . The sparsity of official Soviet information has served to underline one of the key differences between the Soviet Union, where the mass media are rigidly controlled, and the pluralist societies of the West" (Dobbs A1). In the opinion of the Western media, the consequences of the cover-up had been devastating for the ethos of the Soviet Union. By attempting to conceal the accident from European and Soviet citizens, Gorbachev was caught in a contradiction of claiming to support a policy of glasnost while keeping silent about the accident. This contradiction suggested that the new progressive image of the Soviet Union was a façade and that the accident was a moment of truth where the real totalitarian face of the Gorbachev regime had been exposed.

The *Times* and the *Post* reporting on the Soviet Union's causes for secrecy and consequences of the cover-up are taken further in op-ed pieces to make ideological arguments at the stases of quality and policy. In these pieces, Soviet secrecy is used as the basis for qualitative arguments about the difference between Western and Eastern values, one of which was that the United States cared about the well-being of Europeans while the Soviet Union was indifferent to their interests. These qualitative arguments are used to challenge policies promoting greater cooperation between the Soviet Union and Western European nations. In the *New York Times* May 4 op-ed "The Fallout's Fallout," for example, William Safire argues that European nations who have considered closer or more neutral relations with the Soviet Union should reconsider based on what the accident has reconfirmed about the Soviet's true values: "The world's professional innocents wonder why the Soviet Government did not come forward immediately with the truth about the disaster warning neighboring nations of radioactivity headed their way. . . . What is the lesson of Chernobyl to the fraying European side of the Western alliance? This: remember who your friend is; remember which superpower defends your values and which subordinates human life to the power of the state" (Safire A19). A similar, equally negative critique of the Soviets as indifferent to their European neighbors appears on the same day in the *Washington Post* penned by the conservative columnist George Will. In his op-ed "Mendacity as Usual" he writes, "If the wind had not blown radioactivity over Sweden . . . and others, the accident would have remained an Orwellian nonevent. The good for the West that is blowing in this wind is a stark reminder of the kind of regime that runs the Soviet Union. . . . First the Soviet regime

jeopardized neighboring nations by not notifying them; then it began announcing grudging partial quarter-truths. . . . The fire at Chernobyl illuminated a fundamental fact of Soviet culture. Throughout its dark history the
Soviet regime has been willing, even eager, to trade human lives for forced-
draft economic development" (Will C8).

The news reporting and op-ed writing in the *New York Times* and the
Washington Post suggests that the narratives of Soviet secrecy and indifference reoccur and that these themes are ideological, rhetorical arguments of
value and policy, which are meant to extoll American virtue and persuade
European countries to strengthen their alliance with the United States rather
than develop closer relations with the Soviet Union.

An investigation of the visual representations of risk in the media coverage of the Chernobyl accident suggests a new strategy for visualizing radiation risk emerged—the radiation cloud, which supported these themes.
Radiation-cloud visualizations used shaded arrows or stippling over the surface of a map to represent the hypothesized path of fallout from Chernobyl
based on the prevailing wind patterns over Europe. Maps that included these
kinds of visualizations appeared five times in reporting[7] in a date range from
April 30th to May 16th. The use of fallout cloud visuals for representing risk
is illustrated in a map in the *New York Times* accompanying the story "Winds
Blow Fallout to Southern Europe." On this map the cloud image is used by
the paper to communicate both the distribution of the fallout from the Chernobyl accident and the countries affected by it. Stippling, or black dots, show
that the area covered by the cloud from Chernobyl on May 1 stretched hundreds of miles west from the crippled reactor to the eastern border of France
(see fig. 3). Unlike the bull's-eye overlay, the cloud visual communicates the
contours of the area affected by risk suggesting an ovular ring of radiation
spreading from east to west. The concentration of dots in the cloud represents the density of the fallout revealing significantly lower concentrations
of radioactive particulates over Western Europe than in the east.

The radiation cloud visual offers readers of the *Times* more information
about the location and dynamics of the radioactive fallout. However, like most
bull's-eye overlay maps the magnitude of radiation in the areas under the
fallout cloud is conspicuously missing. In the adjacent *Times* article "Winds
Blow Fallout to Southern Europe," however, this information is available and
offers some clarification about radiation levels. The text of the story gives
reassurances, for example, that radiation levels in most of Europe are extremely low: "in Western Europe, the radioactive substances that make up
the cloud are at such low concentrations they are near the limit of detect-
ability" (Browne A8).

It also offers numerically specific statements about radiation levels: "The
intensity of atmospheric radiation reached about 2 millirems per hour early

Figure 3. Path of airborne radiation following Chernobyl. (Reproduction of a map from the *New York Times*; see "Path of Airborne Radiation.")

this week before subsiding. By comparison a passenger flying from Los Angeles to New York receives a dose of 2.5 millirems" (A8).

Like the *New York Times*, the *Washington Post* also uses visualizations of Chernobyl's radioactive cloud that more precisely locate radioactive fallout, approximate its density, and comment on its risk. A map visualizing Chernobyl's fallout, for example, appears in conjunction with the story "Soviets Say Cleanup Under Way" on May 1 (Drew A34). Unlike the radiation map in the *Times*, the *Post*'s map uses arrows to identify the path of the radiation (see fig. 4). These arrows add information about the directionality of the fallout but provide less detail about the presumed shape of the fallout cloud. In order to compensate for the arrows' lack of specificity in defining the geographic areas at risk, the names of countries most affected by the fallout are bolded on the map. Like the cloud visualization in the *Times*, the arrows employ shading to communicate the declining density of airborne radioactive

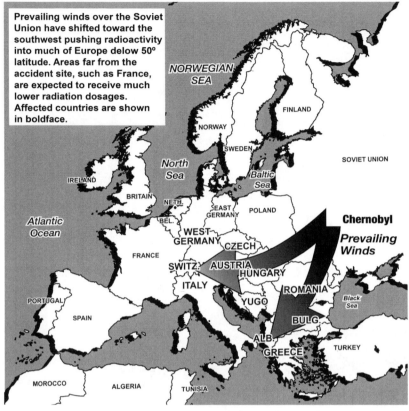

Prevailing winds over the Soviet Union have shifted toward the southwest pushing radioactivity into much of Europe delow 50° latitude. Areas far from the accident site, such as France, are expected to receive much lower radiation dosages. Affected countries are shown in boldface.

Figure 4. Arrow radiation cloud map. (Reproduction of a map from the *Washington Post*; see Drew.)

particles as the cloud moves away from the site of the accident. Also like the *Times*, quantified radiation measurements are not included on the map, but some details about the risk are discussed in an adjacent story. For example, the accompanying article "Some Say Cleanup Underway" explains, "Wind shifts over Europe began bringing clouds of radioactive contamination, which earlier had affected Scandinavia, farther south today, with West Germany, Austria, Switzerland and Italy reporting radiation levels that were abnormally high but not considered harmful" (Lee A1).

From a communicative perspective, the emergence of radiation cloud maps as a prominent visual strategy in both the *New York Times* and *Washington Post* seems explainable because of the advantage it provides in describing the exact location of radiation risk. With the risk of radiation extremely slight in all areas except those immediately around the plant, however, it is important to ask, Why go to such great pains to develop a detailed representation of the fallout and its path if its risk consequences are not all that serious? It could have been the case that the editors of the *New York Times* and *Washington Post* assumed their readers wanted to know exactly where the risk was no matter how slight and inconsequential it might be. Interestingly, such an assumption didn't influence their coverage of the Three Mile Island accident even though it was local, and they knew weather forecasts could be used to predict the movement of radioactive material from the plant.[8] It is also possible that the editors of the *Times* and the *Post* were attempting to serve the interests of their international readers in Europe. However, one has to wonder whether visualizations of minimal risk unaccompanied by more detailed accounts of radiation levels would educate or reassure this audience. Finally, it could be that the editors were trying to grab readers' attention by introducing a novel strategy for representing radiation risk. A review of both government and media sources, however, suggests that the cloud visual was not a novel risk representation. It appeared, for instance, in the film *Radiological Defense* (1961) and the Defense Department's handbook on nuclear attack *Fallout Protection*[9] (1961). It even appeared in *Harper's* coverage of the Three Mile Island accident.[10]

From a communicative perspective, there do not seem to be compelling reasons to explain the rise in prominence of this new strategy for risk reporting. If we consider the choice from a rhetorical perspective, however, the benefits of the increased use of cloud visuals are more obvious. The most compelling exigence would be, of course, geopolitical. The analysis of the textual reporting in the *New York Times* and the *Washington Post* that opened this section offered evidence that these publications had developed ideological narratives about Soviet secrecy and indifference. The analysis that follows suggests that radiation cloud graphics played an important role in reinforcing these narratives. This role is evidenced by the juxtaposition and coordination of reporting discussing Soviet secrecy and indifference with cloud visualizations and by the analogies created in the press between the physical cloud and these character traits.

The capacity of texts and visuals to collectively contribute to frame building in news reporting has been discussed in mass communication literature.[11] One of the ways in which this frame building happens is through the juxtaposition and coordination between the texts of news stories and the visual on the page. The text of the *Times* article with the stippled cloud map, for example, concludes with a statement of outrage about the risks that Western European countries have endured as a consequence of Soviet secrecy and technological incompetence. Quoting "one scientist," the final line of the article states that "not to have warned the West of approaching fallout was absolutely inexcusable. It was bad enough to have built a plant of that type with no containment structure to confine the fallout" (Browne A8). The story also includes the names of countries at risk including Yugoslavia, Greece, Romania, France, West Germany, Switzerland, and Austria (Browne A8). This textual information in isolation makes the argument of Soviet secrecy and indifference and provides exemplars of its victims. What the text cannot do very well, however, is make the abstract risk of radiation concrete for the reader. The cloud visualization, therefore, contributes to the discourse about risk by making the invisible cloud of radiation visible and reinforcing through the visualization the extent and location of the risk.

In addition to making the risk palpable, cloud visuals also contribute to reinforcing the seriousness of the risk. The text of the article is very clear about the low level of risk, and the graphic accompanying the story uses stippling in an effort to communicate this as well. However, the ominous cloud of black dots spreading across the page reminds readers that risk exists. This presencing of the risk in the visualization seems to work against a more sophisticated gradation of radiation risk presented in the text of the article. The visual suggests that no matter how vanishingly small risk is Western European countries are still "under the cloud." Considering the risk in this way has important rhetorical implications. If the Western media wants to create a sense of outrage against the Soviets among Western European nations using narratives of secrecy and indifference, then the risks from the accident must seem tangible and serious. The cloud visualization of radiation supports these narratives by reinforcing the textual arguments through visual repetition and by maintaining the intensity of risk, which is a seminal warrant supporting the unethicalness of Soviet secrecy and the seriousness of their indifference. This could likely not have been accomplished with the bull's-eye overlay whose generality would not have been suitable either for reifying risk or locating it over specific locations. The juxtaposition of radiation cloud visuals with textual narratives of secrecy and indifference as well as the possible value of these visualizations in reinforcing and warranting these narratives, therefore, suggests that cloud visualizations could have been selected strategically to meet the rhetorical exigencies created by the accident.

Whereas the use of cloud visuals by the *Post* and *Times* can be attributed to their practical value in reinforcing and warranting arguments about the existence of risk, they can also be tied to the narratives of Soviet secrecy and indifference. The radioactive cloud, for example, appears in news reporting as a symbol of both Russian secrecy and its revelation. The first and perhaps most obvious analogical association of this kind appeared in the April 30 *New York Times* op-ed piece "Chernobyl's Other Cloud" a day before the first radiation cloud map. The title of the op-ed suggests an analogical connection between the physical radioactive cloud and the conceptual cloud of fear and a cloud of secrecy. The piece argues that it is incumbent upon the Soviet Union to come clean about the accident as an act of good faith toward its Western neighbors. The anonymous author writes "For the Russians to have stayed silent about the disaster for three days does not invite their neighbor's trust. To dispel the *cloud of fear* that has spread beyond its borders, the Soviet Union needs to share promptly all that it knows" [emphasis mine] ("Chernobyl's Other" A30). In these lines, the fear of radioactive fallout from the accident is analogically mapped onto the physical movement of the radioactive cloud. As the latter spreads physically across Europe the former diffuses psychologically as well.

In addition to connecting the physical risk and the psychological fear of the radioactive cloud, the cloud is also used analogically to represent secrecy. In the final line of the same op-ed, the author concludes, "With the same decency, the Russians should hasten to lift *the information cloud* that still hangs over Chernobyl" [emphasis mine] (A30). Unlike the initial analogy, which created correspondences between physical and psychological phenomena, this second one draws on a common reciprocal association between opacity and understanding: the more opaque something is the less it is understood and vice versa. The analogical use of the cloud to represent secrecy surrounding the information of the accident seems to have an interesting and somewhat counterintuitive relationship to the role that visualization of the cloud played in reporting. While the cloud in the textual/analogical sense is a symbol of Russian secrecy, the ability to detect and visually represent the cloud stands for the opposite: a belief in truth seeking and capacity for revealing it. This twist is hinted at in the *Washington Post* op-ed "Chernobyl: Half Hidden Disaster" in which the anonymous author writes, "Taken together with *the readings of airborne radioactivity* in neighboring countries, these signals of distress *provide a far more accurate sense of the events* than the Soviets' tight lipped unhelpful statements" [emphasis mine] (A22). Similarly, other authors commend the Swedes for their technological vigilance that helped expose the accident: "The Soviets owed their neighbors downwind prompt warning of the disaster. Instead, characteristically, they said nothing until the Swedes, 800 miles away began picking up evidence of it" ("Meltdown" A24). As this discourse suggests, the cloud seems to have been coopted, on the one hand, as an analog for the secrecy of the Russians, while on the other, as a representation of the West's (both Europe's and the United States's) efforts to uncover the truth about the accident. The presence of these analogical pairings between the physical cloud and secrecy or clarity suggests that news content creators, and perhaps even members of the news-reading public, had made a conceptual connection between the physical radiation cloud and these abstract concepts. The existence of this connection could have encouraged news producers to adopt cloud visualizations as strategies for making their rhetorical point about Soviet secrecy or Western truth finding, or at least have some confidence that the connection would resonate with their readers.

Just as analogies were used in media reporting on Chernobyl to encourage associations between the cloud and secrecy so too were they used to create conceptual connections with the indifference of the Soviets. In these analogies the cloud is emblematic not only of the Soviet indifference to the safety of its neighbors but also of the political consequences of this indifference. A use of the cloud to illustrate Soviet indifference, for example, appears in the *New York Times* op-ed titled "Moscow's Nuclear Cynicism." In this piece the author Flora Lewis invokes the cloud by quoting a south German newspaper: "It was the militaristic urge to *disregard civilian needs* that led to negligence in setting nuclear standards. '*This is just as much our problem as the radioactive cloud* over Sweden' commented the Suddeutsche Zeitung" [emphasis mine] (A27). Here the use of the line from the German paper about the cloud in juxtaposition with the discussion of Soviet "disregard for civilian needs" suggests that, like the physical radioactive cloud that hangs over Sweden, the cloud of the Soviet leadership's indifference toward the effects of its risk-taking on others hangs over the rest of Western Europe, threatening every country in its path. In this instance the cloud seems to represent the risk of indifference itself.

In other cases, however, the cloud is used analogically to represent the consequences of this indifference. In the *Washington Post* op-ed "A Blow to Nuclear Arrogance" by Stephen Rosenfeld, for example, the author writes, "Here [in the case of Chernobyl] there are deaths, *a plume of radioactive poison* drifting over hundreds and hundreds of miles of settled land and across national frontiers, and *a matching plume of immense medical, economic, and political consequence*, especially for Moscow" [emphasis mine] (A19). In this line, the author creates an analogy between the movement of the radioactive cloud over Western Europe and the spread of the political fallout in the region for the Soviets because of their indifference toward risk management. Like analogical representations of secrecy, the representations of indifference extend the concrete properties of the cloud and its movement into the abstract ideological domain. Because both of these associations appeared before and during the period in which cloud visualizations became primary features of the visual landscape in news reporting, it seems reasonable to suggest that they might have encouraged or supported the use and interpretation of cloud visuals as representations of Soviet indifference and secrecy.

A detailed examination of the text and visuals in the *New York Times* and *Washington Post* used in reporting on Chernobyl suggests that cloud visualizations emerged as a response to the political exigencies of the nuclear accident. An assessment of the political context, the juxtaposition of textual and visual information, and the presence of analogical connections illuminates the existence and circulation of a relationship between the radiation cloud and narratives of secrecy and indifference. This visual/verbal relationship played an important rhetorical role in news reporting on the Chernobyl accident by reinforcing the presence of risk, its general geographic extent, and its seriousness for Western Europe. These contributions helped underwrite ideological claims about American moral superiority and broadcast to European nations the consequences of developing close alliances with the Soviet Union.

Risk Representation in the Age of the Internet: The Rise of Citizen Science and the Birth of Grassroots Risk Visualizations

The preceding sections have introduced the major conventions for risk reporting in the mainstream media on nuclear plant accidents prior to the digital age. These explorations have shown that in the predigital era bull's-eye overlays and cloud visualizations dominated the visual landscape and that their dominance was informed both by the communicative limitations and ideological and practical exigencies of their creators. With the advent of the accident at Fukushima, a new and important chapter of risk representation opened. Unlike previous nuclear accidents, Fukushima was the first to occur in the age of the Internet and the first for which there were citizen-led initiatives to create and publicize visual representations of risk. This section explores how the Internet and the emergence of citizen science have influenced the style and content of risk visualization. By comparing the risk representations created by the mainstream media with those developed by the grassroots citizen-science group Safecast, this section provides insight into how changes in the source of visualizations can influence the manner in which risk is communicated. In so doing, it endeavors to illuminate the particular characteristics of grassroots citizen-science risk visualization and account for the differences between these visualizations and the conventional strategies used by the mainstream media. This examination suggest that whereas risk representations created by the mainstream media tend to generalize risk and its distribution, grassroots risk representations offer precise and comprehensive accounts of radiation magnitude, type, and distribution. These differences in risk representation seem to be consequences of divergences in communicative goals and audiences between the mainstream media and community-centered risk reporting.

Fukushima

At 2:46 p.m. Japanese Standard Time (JST) on Friday, March 11, 2011, the 9.0 magnitude Great East Japan Earthquake struck the island of Japan, wiping out power and wreaking havoc on its infrastructure. At the moment of the earthquake, all three of the reactors online[12] at Fukushima Daiichi went into emergency shutdown, and backup diesel generators seamlessly main-

tained the operations of their cooling pumps after the electricity to the plant
was cut off (IAEA). Approximately forty-five minutes later, however, the first
tsunami wave struck the facility, overtopping its protective seawall and disabling the cooling exchanges at the ocean's edge. Minutes later a second larger
wave hit, disabling twelve of the plant's thirteen backup diesel generators
(World Nuclear Association 1).

Without electricity to run the pumps bringing cooling water to the reactors, only one line of defense remained: the reactors' emergency core cooling system (ECCS). The ECCS is a set of interrelated safety systems designed
to protect the reactor core from overheating when primary systems for core
cooling fail by injecting water into the reactor core and managing the amount
of pressure in the reactor vessel. These systems functioned for almost an
hour after the tsunami and then failed (World Nuclear Association 6). In
less than four hours, the core of Unit 1 had completely melted and breeched
the reactor pressure vessel. Three days later on March 14, 2011, after her-
culean efforts by firefighters and plant workers to control temperatures in
units two and three failed, their cores also melted and breeched their pressure vessels (Shimbun).

In the days following the earthquake, tsunami, and nuclear meltdown,
the *New York Times* and *Washington Post* scrambled to keep readers updated
on the details of the multiple disasters. As the initial shock of the tsunami's
massive devastation subsided, the unfolding drama of the nuclear meltdown
at Fukushima drew increased attention, particularly after the explosion in
Unit 1 on March 12. From March 13 to March 20, the *New York Times* and
the *Washington Post* provided substantial coverage of the nuclear disaster,
which included radiation risk visualizations. Generally, the strategies for visualizing radiation risk in reporting on the Fukushima accident were similar
to those used in previous accidents by the *Times* and *Post*: Bull's-eye overlays
and cloud visuals dominated.[13]

Whereas the type of visual conventions for reporting radiation risk remained
largely unchanged, the style and content of these strategies were transformed
by differences in the technological and informational circumstances of the
accident. Perhaps the most significant technological change in risk reporting
was the advent of the Internet whose affordances for presenting information
in rich multilayered formats known as "mashups" impacted the style for visualizing risk in print-news formats. The defining characteristic of online
mashups is that they are aggregates of discrete nodes of information, which
are thematically coherent but separated from each other in cyberspace. These
nodes are conceptually bound together by an anchor, a text or visual that a
user would encounter first while browsing, which provides a hub for accessing
offscreen nodes. By clicking links embedded in the anchor, offscreen nodes
appear overlaid on the text or visual or in a separate window. These features

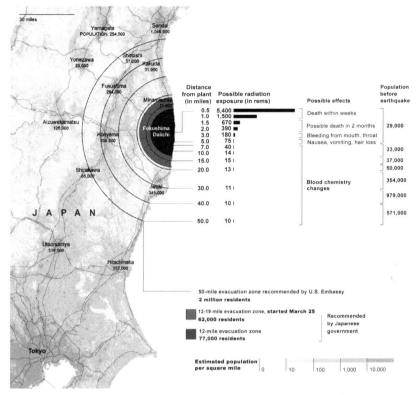

Figure 5. Media mashup map of radiation. (Reproduction of a map from the *New York Times*; see Cox, Ericson, and Tse A11.)

offer solutions to the problem of information density by allowing individual layers of information to be peeled off and sequestered in cyberspace. At the same time, links in the mashup connect the informational nodes to the anchor and to each other promoting connectivity and facilitating ease of access.

The influence of online mashups on print visualizations of radiation risk following the accident at Fukushima is illustrated in the graphic associated with the *New York Times* story "Data Show Radiation Spread; Frantic Repairs Go On," which appeared on March 18, 2011 (see fig. 5) (Cox, Ericson, and Tse A11). A glance at the graphic reveals that it is divided into two visual columns. The left column is dominated by a giant map of Japan with a bull's-eye overlay while the right column is subdivided into three sections by two horizontal lines.

Of all the elements in the graphic space, the map of Japan commands the most attention. Its size and position at the beginning of the reader's left-to-right scanning path suggests that this element is the print equivalent of the anchor in an online mashup. Like the radiation visuals of Three Mile Island and Chernobyl, this representation includes the source of the risk, its geographical extent, and the populations affected by the risk. However, unlike previous print visualizations, the visual space for risk representation has been expanded to include additional risk information. This additional information is organized using the rule of thirds, which is typical for print and website design where titles are presented at the top of a page, content in the middle, and auxiliary information in the page footer (Lynch and Horton 89). Further, the content in the center of the page emulates the format of online mashups by presenting individual nodes of information and relating it to the visual anchor. Each of these distinct nodes includes information about a different dimension of the risk situation including distance from the risk event, the magnitude of radiation risk, the possible effects of radiation, and the population living in the area affected by the risk. Although each of these nodes is distinct, they are linked to one another and the anchor using a variety of visual strategies.

In most instances the information presented in these nodes is the same kind presented in previous print maps. However, the amount and variety of information is increased and the information is separate from rather than overlaid on the visual. Of greatest interest are the radiation measurements, a content feature that never appeared in the reporting of the *New York Times* or the *Washington Post* during the Three Mile Island and Chernobyl accidents. In reporting on the Fukushima accident, radiation measurements appeared in risk visuals on two separate occasions, both of which were in the *New York Times*.[14] The appearance of measurements in these instances is related to a dispute between the US embassy and the Japanese government over the size of the evacuation zone. For the sake of brevity, the strategic uses of these measurements will not be examined here. Instead, it will suffice to note that unlike in the Three Mile Island and Chernobyl accidents, the radiation levels around Japan following the Fukushima accident were more comprehensively measured and widely available. In the case of the map accompanying the article described in the previous paragraph, these values were drawn from measurements taken and reported by the US Department of Energy's National Nuclear Security Administration. In other reporting, the *New York Times* used radiation levels from the Japanese Ministry of Education, Culture, Sports, Science, and Technology's (MEXT's) website Mext.go.jp ("Japan's Assessment"). The availability of this information, along with the political exigence to use it in the debate over the evacuation zone, suggests why it might have appeared in these visualizations but had been absent in previous reporting.

This brief visual assessment of a sample of risk representation in the *New York Times* suggests that the style and content of print media had been influenced by online formats, particularly the media mashup. This movement of new digital formats into traditional print style seems to run counter to the process of remediation outlined by Bolter and Grusin (Bolter and Grusin 45). Instead of recycling or repurposing old media, what seems to be happening here is a kind of retrocycling or retropurposing, where traditional media are being "modernized" or updated by adding stylistic elements from new media. Though modernization may be one reason for retropurposing, a more pragmatic reason might be the requirements of producing news simultaneously in both print and online formats. Like most other major newspapers, the *New York Times* produces both print and online versions of their paper. Because they are working in two formats, it makes sense from a production standpoint to create visuals that can function in both rather than producing different visuals for each. In fact, the visualization previously discussed appeared in both the print and online versions of the *New York Times* in virtually the same format (Cox, Ericson, and Tse A11; Cox, Ericson, and Tse, "The Evacuation Zones"). Considering the requirements of news production, it seems that this visual representation of risk is a hybrid style emulating the format of a media mashup but adapting it to the limits of the printed page.

In addition to producing hybrid visualizations for the printed page, the *New York Times* and *Washington Post* also developed static infographics and interactive maps that were exclusively associated with their online reporting. Under the title "Japan's Nuclear Emergency" *washingtonpost.com* published a series of static infographics, some of which appeared in its print reporting and some of which were exclusive to the Internet. Interestingly, though these infographics covered a variety of topics connected with radiation risk,[15] there were no efforts by the *Post* in their online reporting to visualize the location or magnitude of radiation risk from the accident. These types of visualizations only occurred in their print reporting and even then only included information about the hypothetical spread of the radiation plume as it moved across the Pacific toward the United States.[16] In contrast, the *NYTimes.com* provided both static print maps of radiation risk and an interactive one. The interactive map was included as part of a suite of interactive maps generally titled "Map of the Damage from the Japanese Earthquake." By selecting "radiation levels" on the infographic's navigation bar, the user could bring up a map of the area within a 50 kilometer radius of the accident. Within this range, the 20 kilometer evacuation zone and 30 kilometer stay-inside zones instituted by the Japanese government are marked with a bull's-eye overlay (see fig. 6). By zooming out slightly, the user can also see the 50 km radius of the evacuation zone suggested by the US embassy marked in a circle of red dots on the map. Between the 50 kilometer and 20 kilometer zones, the map includes 44 dots ranging in color from white to dark purple, which indicate increasingly higher levels of radiation (Bloch et al.). When the user's

Map of the Damage From the Japanese Earthquake

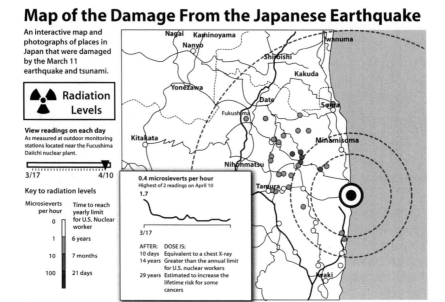

Figure 6. Online map of Fukushima radiation risk. (Reproduction of a map from the *New York Times*; see Bloch et al.)

cursor moves over a dot, a text box appears on the page with the radiation level measured in microsieverts per hour (μSv/h), the date of the measurement, the number of measurements taken, a graph of the change in measurement over time (between March 17 and April 10, 2011), and a comparative contextualization of the measurement. The comparative contextualization includes two columns. The left column supplies a quantity of hours, days, months, or years, and the right contains one of three radiation dose or risk comparatives: "equivalent to a chest x-ray," "greater than the annual limit for U.S. nuclear workers," or "estimated to increase the lifetime risk of some cancers." Jointly, the columns are supposed to help the reader understand how at risk an individual in the indicated spot on the map would be. Someone in a spot measuring 7.3μSv/h, for example, would have a dose equivalent to an x-ray in 14 hours, would exceed the annual limit for nuclear workers in 285 days and increase their lifetime risk of cancer in 2 years (Bloch et al.).

By providing an interactive map that includes specific information about radiation levels near the accident, the *New York Times* allowed their online readership in both the United States and Asia to explore the risks of radiation from the accident in more detail than they had previously been able to do thanks to the affordances of the Internet and the availability of information about radiation risk. Despite this level and type of detail, these maps were still qualitatively and quantitatively limited when compared to visualizations created by the citizen-science group Safecast. This disparity I will argue is a consequence of differences between the goal, audience, and information-gathering practices of the mainstream media and those of the citizen scientists in Safecast.

Visualizing Risk: RDTN.org and the Development
of Online Radiation Mapping

The visual conventions of the mainstream media for representing radiation risk, both in print and online, and the contextual factors that influence these choices provide a point of reference from which we can begin to appreciate the unique character of risk visualizations developed by grassroots citizen-science risk visualizers. This section identifies and accounts for the differences between radiation risk visualizations created by the citizen-science group Safecast and the visualization practices of the mainstream media. To make this comparison, I have divided Safecast's efforts into two stages. In the first stage "risk visualization," the organization, which was initially called RDTN.org was involved primarily in collecting radiation measurements from existing institutional sources and visually representing them in an online map. In the second stage, "establishing a sensor network," RDTN.org had formally become Safecast.org and had embarked on an effort to collect its own data on radiation risk using citizen-hacked radiation sensor technology. By following the details of this transformation and the interplay between data gathering and risk visualization, the following sections chronicle the emergence of a grassroots citizen-science effort to produce visualizations of radiation risk that had never before existed.

The first phase of Safecast's development took place roughly in the period between the launch of the website RDTN.org on March 19, 2011, and the group's renaming as Safecast on April 24, 2011. In this initial period, the group that would become Safecast materialized organically from a loose confederation of designers, hackers, and online entrepreneurs, all of whom had become concerned about what the radiological risks of the Fukushima accident meant for themselves and for their friends and family living in Japan. This grassroots confederation began to coalesce with the launch of the website RDTN.org, which was conceived by Portland, Oregon, designer Marcelino Alvarez and developed with the help of his colleagues at Uncorked Studios. According to Alvarez the creation of the website was initially a selfish endeavor that grew out of his desire to know whether an event happening on the other side of the Pacific Ocean was something that he should be worried about living on the West Coast of the United States (Alvarez, interview). In the days following the accident, he had initially turned to the American mainstream media for information about the magnitude and spread of the radiation risk but became increasingly frustrated with its offerings. Spurred on by this frustration he decided to build a website to aggregate and visualize information about the radiation levels in Japan. In his personal blog at Uncorked Studios on March 21, 2011, he explains, "I wanted to gather all the information that the talking heads did not have time to talk about. I wanted to create a site so simple and easy to use that it would allow anyone the ability to check and see if all things were clear" (Alvarez, "72 Hours"). In addition to personal motives, Alvarez was also driven by a sense of democratic idealism that his website could provide information for and by the people about radiation risk: "We thought there was something noble to the notion of having people purchase their own detection devices and post data" (Alvarez, "72 Hours"). Energized by these frustrations and convictions, he and his coworkers developed and launched the website RDTN.org on March 19, 2011, a little more than a week after the accident at Fukushima.

Though the website was imagined as a space for the average person to post their data on radiation from their own detection devices, in its very earliest inception RDTN.org included very few noninstitutional radiation measurements. In fact, the bulk of the data on the site (roughly 80%)[17] was scraped from the Japanese Ministry of Education, Culture, Sports, Science, and Technology's (MEXT) website (Zhang; James; Aaron). The information that did not come directly from MEXT was drawn from a mix of institutional and noninstitutional sources that had downloaded their data to Pachube,[18] a web-based service that managed real-time data from web-connectable devices. Though Safecast didn't offer substantially more information than could be gleaned from institutional sources, it could provide something that institutional sources did not, an interactive visualization of radiation levels in Japan. Unlike MEXT's website, for example, where users had to read through a densely packed spreadsheet ordered by prefecture to find a radiation level reading, RDTN.org made identifying the radiation levels easy: find your desired location on the map and click.

When compared with the visualizations in the mainstream media like the *New York Times* interactive map, the Safecast map, while equally user-friendly, provided a more comprehensive and nuanced account of radiation risk. In its earliest stage of visualization, the radiation data collected by RDTN.org was presented in a typical Google maps GIS overlay format (see fig. 7) in which the locations of radiation measuring devices were marked with standard Google map location icons. It provided users with radiation readings from more than one hundred individual measurement sources, more than double the number of locations offered by the *Times*'s interactive map. In addition to being more comprehensive, RDTN.org was also more nuanced in some of the information it supplied about radiation measurement. Like the *New York Times* interactive map, clicking on an icon opened a window,

測定者： 田村裕和 Tamura Hirokazu 東北大学理学研究科物理学専攻
Department of Physics, Tohoku University Graduate School of Science

Current reading: 0.18 μSv/h (microSievert per hour)
Time of reading: 計測日: 2011/03/22 05:25:00
Equipment used: Survey meter y ALOKA TCS171 (NaI counter)

Figure 7. RDTN radiation map of Japan. (Used by permission from Marcelino Alvarez; see *RDTN Radiation*.)

which displayed further information about the current reading of radiation at that location in microsieverts per hour (μSv/h) and nanograys per hour (nGyh), as well as the date the reading was taken. Unlike the *Times* map, however, the icon information on RDTN's map also included the location of the source for individual readings. By double clicking on the map the user could also zoom in or out, creating comprehensive views of the number of different measuring stations across the country or local views showing how many measuring devices were taking readings in a particular area. In the *Times* interactive map, spots with different radiation levels were marked. However, no names for the locations were given and the areas on which the spots are marked are devoid of most navigational features such as roadways, mountains, or rivers. Though the *Times* map does contain a few major cities, only a very general sense of location of the measurement sites can be extrapolated from these (Bloch et al.).

In contradistinction, the RDTN map provides a more comprehensive account of radiation risk and a more nuanced description of its location. These

differences, though slight, when considered as communicative choices illustrate how variations of goals and audiences influence risk visualization. The
use of a limited number of radiation sites and the absence of details about
the locations of the sites on the *Times* interactive map can reasonably be attributed to the goals of and audiences for mainstream news reporting. The
goal in the mainstream media is to relate the basic facts about the accident
in an accessible and concise fashion. The *Times's* interactive visual goes beyond the basic facts by providing some precise levels of radiation data; however, it is not surprising that the creators of the map stopped far short of
offering users a comprehensive set of data points. Although the print and
online versions of the *Times* circulate in Japan and other parts of Asia, its
primary audience is the American news-reading public. For the majority of
this audience, a few data points in the area around the plant would be sufficient to satisfy most of their demands for information about the radiation
levels. Further, geographically contextualizing these data points with a few
navigational features would be sufficient for American news consumers to
get the gist of the location of the radiation while at the same time avoiding
information overload on the map.

For RDTN, on the other hand, comprehensiveness and detail rather than
sufficiency and concision are the primary design parameters. These goals
are clearly spelled out in their website blog where they write, "Our hope in
launching this site is that clear reliable data can provide *focus* on the critical relief efforts needed in Japan. . . . We have been working day and night
to find and integrate new data sources that can help provide reliable data"
(Ewald). Because the goal of RDTN's citizen-science project was to get reliable data and find new data sources, we can extrapolate that the purpose
for getting this data is so that it can be visualized on the website to provide
a comprehensive map of radiation levels across Japan. Another top priority
would be making those measurements meaningful for Japanese citizens.
These priorities account for why RDTN's online map included precise details
about the locations of the radiation sensors as well as why it included navigational details such as roads, mountains, and rivers. By incorporating these
features, the creators of the map help the user better judge what is known
about the radiation level in their vicinity. Although in this protean format
the measurements may not be as local as users desire, they nonetheless include the maximum amount of information publicly available about radiation risk levels at the time the map was created.

RDTN's development of an online map that provided comprehensive and
detailed data about radiation in an easily accessible format immediately caught
the attention of a small group of well-connected online entrepreneurs who
were either living in Japan or had family and friends who lived there. They
included Sean Bonner, an online-entrepreneur, journalist, and activist in Los

Angeles; Joi Ito, venture capitalist, high-tech entrepreneur, activist, and, at the time, the soon-to-be head of MIT Media lab; and Pieter Franken, an information technology manager in the Tokyo banking sector and visiting researcher at Keio University in Tokyo. In the aftermath of the earthquake and tsunami, Bonner, Ito, and Franken had been in constant contact with each other and with their friends and family in Japan, checking on their safety and inquiring how they might help them through the disaster. With the advent of the nuclear plant crisis, these discussions of aid turned toward the possible risks of the accident and talk of getting Geiger counters to their loved ones in Japan (Bonner, interview). On March 18, the day before the launch of the RDTN, the three were introduced to Marcelino Alvarez and his team in Oregon. Because of their mutual personal interests in understanding the risks associated with the accidents, the two groups agreed to cooperate to develop RDTN.org. In the weeks following their introduction, Alvarez and his colleagues at Uncorked Studios continued to add information and tweak the design of the website while Ito, Bonner, and Franken provided publicity and technical support for it. Bonner produced the first write-up of RDTN in the online magazine *boingboing* helping RDTN get further attention in both online and traditional media. Ito introduced the group to academics and specialists in the fields of physics and computer science that could help them with the development of the website. Collaboratively they began a citizen-science project that would revolutionize radiation risk visualization.

Developing New Visualizations: The Safecast Map

As the US-based RDTN began to develop and get attention in the media for its efforts to visualize risk, a parallel citizen-science endeavor for radiation measuring was starting to take shape across the Pacific in Japan. This endeavor was set in motion by American expatriate Christopher Wang, known as Akiba, who was a member of Tokyo Hackerspace, a club dedicated to modifying existing technologies to extend their utility or create new devices ("What Is Tokyo Hackerspace?"). On March 15, 2011, the Tuesday following the earthquake, Akiba and his tech-savvy hacker compatriots met at Tokyo Hackerspace to decide how they could use their technical skills to respond to the crisis created by the earthquake, tsunami, and nuclear accident. They decided to build solar lanterns for local residents who were without power and to develop a Geiger counter network to collect radiation readings in Tokyo and then expand the network out into the rest of Japan. The decision to build the network was driven by the personal needs of the group to know what the radiation risks were for them and by a collective skepticism about the Japanese government's willingness to be transparent about the risk. In a blog post about the meeting on March 16, Akiba writes, "The real worry for many people here is the lingering effects of radiation. The public does not trust the gov-

ernment numbers very much at the moment. . . . This [Geiger counter] project benefits all volunteers and pretty much anyone who falls in the range of
the network" ("Thanks").

Because of his experience building custom electronic devices,[19] Akiba assumed the task of hacking or fabricating Geiger counters for the group's sensor network. His initial efforts were dedicated to modifying one of two do-
nated Cold War–era Geiger counters so that its radiation readings could be
downloaded to the web. He also had plans for fabricating Geiger counters
from scratch using homemade circuit boards and radiation tubes ordered
from the Ukraine (interview). These projects were extremely important, because there were few Geiger counters available in Japan at the time of the accident and no commercially available Geiger counters that could upload radiation data to the web. This latter issue was significant because the group
wanted their radiation readings to benefit Tokyo residents broadly and to
be a source for checking official radiation readings. By March 23 Akiba had
successfully hacked a Cold War Geiger counter and began broadcasting radiation readings taken outside his apartment to the web (Akiba, "Hacking").
With this success came attention. John Baichtal, a writer at *Make*—an online magazine dedicated to reporting the news and projects in the hacking
community—found out about the hack and wrote a short piece on the project
the same day. In the wake of this publicity, Bonner encouraged Franken, who
was living in Tokyo, to visit Tokyo Hackerspace (Bonner, interview). On April
2 Franken, an electronics enthusiast, participated in one of Tokyo Hackerspace's solar lantern builds. At the event he introduced himself to Akiba, and
the two discussed the group's plans to establish a radiation sensor network.

By the end of March and beginning of April, the original RDTN confederation of Alvarez and his colleagues at Uncorked Studios and Ito, Bonner,
and Franken expanded to include Akiba and a few other members of Tokyo
Hackerspace. The participants in this newly expanded confederation recognized that their fundamental problem was the lack of Geiger counters. RDTN
had a website but very little nongovernment data to broadcast. Akiba and Tokyo Hackerspace had cleared some of the technical hurdles associated with
building web-connected Geiger counters but needed resources to expand their
network. Recognizing this mutual need they decided that moving forward
required raising money; so, on April 8, 2011, RDTN and Tokyo Hackerspace
jointly broadcast a pledge for donations to purchase Geiger counters using
the Internet fundraising site Kickstarter (Aaron). They also realized that with
the expansion of the group it was important to work more closely to coordi-
nate their efforts. In response to this exigence, representatives from each part
of the collective agreed to meet for face-to-face discussions about the enterprise at the New Context Conference in Tokyo on April 15, 2011. During the
conference, they made formal plans for the cooperative development of the

sensor network and decided that the confederation should become a single comprehensive entity. According to the official account on Safecast.org, the group decided they needed to "focus on collecting data and concluded that a new brand was needed to describe both the work we were doing now and in the future. . . . We were to be called Safecast" ("History").

By the time the citizen-science collective officially changed its name to Safecast on April 24, 2011, the transformation of its mission from strictly data visualization to data gathering and visualization had already begun. On April 14 the group deployed their first Geiger counter with a volunteer, Dave Kell, who was traveling to Ichinoseki and Sendai north of Fukushima to deliver food for the relief organization Second Harvest. Kell reported data from a Geiger counter in his car on his drive by taking pictures of the probe with his iPhone and then sending the picture to the online photo site Flickr. The photos were then posted to Twitter so that the readings could be followed in real time (Bonner, "First RDTN Sensor"). Although this method of radiation reporting was serviceable, the group wanted to create more efficient means of collecting and uploading radiation measurements to the web. Their most significant innovation was the development of the bGeigie, a bento-box-sized radiation-measuring device that could be strapped to the side of a car and driven around to take and download continuous radiation measurements. The first bGeigie was taken out for a test drive on April 23, 2011, by a team that included volunteers from Keio University and Pieter Franken. Driving north from Tokyo toward Koriyama in the Fukushima prefecture, the team took measurements around elementary, junior high, and high schools (Bonner, "First Safecast"). Because of its mobility and automaticity, the device provided a simple and reasonably inexpensive method for collecting measurements across a large area, a boon for a small grassroots organization aspiring to provide a comprehensive survey of radiation levels.

As collecting radiation measurements became a central part of Safecast's mission, the character of the visualizations they created began to change. Though there are a number of visual innovations that arise out of the group's data-gathering efforts, their signature visualization was the Safecast map. The map was announced on June 23, 2011, and has been updated several times.[20] Perhaps the most remarkable difference between the Safecast map, the maps in the mainstream media, and the early RDTN maps is the level of detail about radiation risk it includes. Whereas RDTN.org's map provided over a hundred measurements in contrast to the *New York Times'* interactive map with tens of measurements, the Safecast map included initially hundreds of thousands of measurements, a number which even as I write has climbed into the millions. The impressive scope of the measurements is thanks to the bGeigie. In the weeks and months following its invention, the bGeigie was driven up and down the east coast of Japan north from Tokyo through

the area around Fukushima and up to Mutsu, one of the northern-most cities on the main island. Volunteers also drove bGeigies south from Tokyo to
Saga, a city at the far-southern end of Japan. With the data Safecast created
a series of maps in which drive routes were covered with colored dots. Each
of the dots indicated a radiation measurement. Cooler colors (green or blue)
represented levels of radiation either commensurate with or barely elevated
above levels of radiation previous to the accident, and hotter colors (red or
yellow) indicated radiation levels that were substantially higher.[21] When a user's cursor was placed over each individual dot a text box appeared, which included the precise level of radiation measured in counts per minute (CPM)
and microsieverts per hour (μSv/h), the GPS coordinates of the measurement, and the date on which it was taken.

The use of color to represent different levels of radiation risk was not a
Safecast innovation. The *New York Times'* interactive map employed a color
scheme for differentiating risk levels. Color scaling of radiation risk also appeared in maps created by the US Department of Energy's National Nuclear
Security Administration and by nongovernment organizations like Greenpeace (*Aerial Monitoring; Map of Radiation*). What was unique about the Safecast map, however, was that whereas the maps from other sources used color
to give a general sense of radiation levels, the creators of the Safecast map
wanted to use it to describe radiation risk at both general and local levels.
As a self-described citizen-science endeavor, part of Safecast's mission, like
RDTN's, was to provide Japanese citizens with as much local information as
they could about radiation risk. Members of the group reference this goal
in interviews following the rebranding of the group as Safecast. In an interview with *Al Jazeera*, for example, Sean Bonner explains, "Despite the alarm
inside Japan and abroad, specific information about radiation levels and its
range are still mostly unavailable. This lack of information is what Safecast
is trying to overcome" (Jamail). This goal of providing maximum information about risk is also evident in the inclusion of the GPS coordinates of the
measurements on the map that allow Japanese citizens with GPS devices to
identify the risk levels nearest to their location.

Whereas specific detail is necessary to give Japanese citizens a more precise understanding of their particular risk situation, geographically broader
representations of risk are also useful. What is notable about the range of
risk from a nuclear accident is that it can vary from street to street and town
to town. Radioactive hot zones can appear in isolated pockets many miles
away from the accident and cool zones can pop up in the immediate vicinity
of the reactor. These inconsistencies reveal the capriciousness of the forces
of weather and topography that govern the distribution of fallout. As Safecast volunteer Kalin Kozhuharov explains, a comprehensive view of radiation
risk is required for the Japanese public to fully appreciate the nuances of the

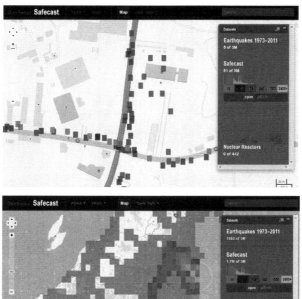

Figure 8. Safecast map subordinate and superordinate views.
(Used by permission from Safecast; see *Safecast Map*.)

distribution of radiation risk from the Fukushima accident: "There is a need
for this [radiation sensor] network because wind and earthquakes can change
radiation levels. 'The data needs to be organized well so people can get the
whole picture of the situation'" (Watanabe). The challenge for Safecast was
to consider how to provide its users with both the local information they desired as well as a broad geographic representation of risk that would let them
see the idiosyncrasies of the distribution of radioactive fallout.

To solve this problem, Safecast adopted a grid map for representing the
data. This visualization technique involves layering risk information in a set
of nested grids in which the whole grid at a subordinate level becomes a piece
of the grid at a superordinate level (see fig. 8). Users move up and down these
levels by zooming in and out of the map. At its most granular level, the grid
of the Safecast map shows an area of several hundred square meters. In this
space the dots or squares representing individual measurements appear. As
the user zooms out a few levels the individual measurements are replaced

with a grid of larger colored squares whose colors and radiation levels are determined by averaging all of the dots and squares at the subordinate level the larger square encompasses. By zooming out even further, the user can examine the distribution of radiation risk around the whole region and, at the highest resolution, the whole country. This strategy for representation allows users, for example, to identify a hotspot twenty miles northwest of Tokyo at the regional level that at a higher resolution can be pinpointed to a stretch of highway 2 kilometers west of Daiichi Hospital between the cities of Ota and Isesaki (*Safecast Map*).

The Safecast grid map avoids the weaknesses of the conventional bull's-eye overlay and radiation cloud visualizations but incorporates their strengths. Like the bull's-eye overlay, the Safecast grid map at the very broadest level of visualization allows the user to have a sense of the average levels of radiation in areas around the plant. In fact, the map even incorporates a very basic bull's-eye overlay design which outlines the 20 kilometer evacuation zone around Fukushima and in earlier iterations also included the 50 kilometer ring marking the evacuation zone suggested by the US Embassy. Because the visualization is built up dynamically by continually averaging radiation values from more local measurements rather than statically using the flat average of all measurements in the area circumscribed by the concentric ring, the grid map avoids the bull's-eye overlay's inability to represent the variability in the distribution of radiation risk from a nuclear accident. Interestingly, in talking with the media about their radiation risk visualizations, members of Safecast reveal a keen awareness of the value in their technique for overcoming the limitations of the bull's-eye overlay. In an interview in fall 2011 with PBS NewsHour correspondent Miles O'Brien, Safecast's Sean Bonner highlights the superiority of their visualization by contrasting it with the shortcomings of the conventional bull's-eye overlay:

> Miles O'Brien: Sean Bonner is one of the founders of Safecast, an all-volunteer organization that has plotted the most detailed maps of radiation contamination in Japan. . . . Radiation doesn't fit that nice neat little disk they want to paint on the map, right?
> Sean Bonner: Right. Right. Yes. Radiation isn't looking at a compass radiating outward.
> Miles O'Brien: Yes. That's right. It's a very arbitrary thing.
> Sean Bonner: Yes. There is wind and topography and this crazy stuff that ends up playing into it. (O'Brien)

Here O'Brien and Bonner take aim at the bull's-eye overlay's inability to capture the nuances of radiation movement. O'Brien, for example, critiques the overlay design's inherent prejudice toward an orderly symmetrical advance of radiation with the comment/question, "Radiation doesn't fit that nice neat little disk they want to paint on the map, right?" In addition, the two team up to emphasize the asymmetrical nature of the risk. O'Brien begins the critique with the statement that "It's [radiation is] a very arbitrary thing," and Bonner elaborates by talking about the variety of factors that play into its distribution: "there is wind and topography and this crazy stuff that ends up playing into it." By adopting the grid map design, Safecast is able not only to illuminate the complexity of the distribution of radiation risk but also critique conventional styles of risk representation.

Whereas Safecast's design permits them to overcome the bull's-eye overlay's inability to accurately show risk distribution, it also allows them to address the problems associated with the cloud visualization. Though the cloud visualization provides a clearer account of the distribution and direction of radiation risk, it nonetheless is opaque about the actual magnitude of the risk under the cloud. The Safecast map solves this problem by providing the reader with either exact or averaged radiation measurements for locations selected on the map. This connection of data and information helps to make the level of risk more transparent for the user. Perhaps more remarkable, however, is the number and types of radiation levels it reports. While most maps in the mainstream media include radiation levels in a single standard unit of measure (rems, sieverts, or grays), Safecast's visualization reports them using both standard and nonstandard units. The standard unit of microsieverts is always used, but the nonstandard unit counts per minute (CPM) is also employed and appears as the default measurement on Safecast maps. The selection of a nonstandard unit as the default, though unusual, has important implications for the public's understanding of radiation risk.

What is not commonly reported about radiation risk in the mainstream media is that there are three types of radiation that can be associated with a nuclear accident: alpha, beta, and gamma radiation. In every case in which the mainstream media includes specific radiation measures, it represents only the levels of gamma radiation. This type of radiation is considered most serious because it can penetrate skin, clothes, and buildings and increase the risk of cancer and other harmful physiological consequences. Alpha and beta radiation, however, are generally considered less risky because they are less penetrating. Alpha radiation, for instance, can be blocked by paper and beta by clothing. However, these types of radiation can still have substantial health risks. If radioactive materials emitting these types of radiation are inhaled or ingested, they intensely irradiate local tissue because their radioactive energy is densely concentrated. Because the probability of ingesting or breathing radioactive particulates was not negligible, particularly for children, the members of Safecast believed that the real radiation risks associated with the accident had been underreported (Bonner, "Alpha, Beta, Gamma"). In

order to provide a more comprehensive assessment of the risk and to educate users on the variety of the different kinds of radiation, the group decided to use CPM as their standard unit for reporting. As Bonner explained
in an interview posted by the technology website O'Reilly Radar, "The sensors Safecast is deploying capture alpha, beta, and gamma radiation. 'It's very
important to track all three'" (Howard). In their reporting of radiation using
CPM, Safecast moves beyond even the most detailed cloud visualization representations of risk to include information about types of radiation not considered significant by the mainstream media. This choice provides further
evidence that the design of their risk representation is guided by a mission,
which diverges from the mainstream media's: to make radiation risk transparent and to educate the public about its complexity.

Conclusion

By examining the different strategies for visualizing radiation risk over time
in the mainstream media as well as the development of novel visualizations
by the citizen-science group Safecast, this chapter has revealed that the digital age has enabled new types of risk representation and that differences
between mass media representations of risk and newly emerging citizen-
science representations reveal that these groups have divergent communication goals. The development of radiation risk visualizations by the mass
media across the accidents at Three Mile Island, Chernobyl, and Fukushima
reveals that choices of risk visualization were influenced by a variety of factors including the availability of information about radiation risk, the technological affordances for reporting risk, and the ideological commitments and
goals of risk reporters. With the rise of the Internet, the capacity for noninstitutional representations of risk became a reality. By engaging intimately
with the science, technology, and methodologies for gathering and reporting
radiation risk, the group Safecast developed novel risk visualizations to serve
their needs and what they imagined were the needs of the Japanese public.
Because they had the opportunity to represent and collect data about risk,
Safecast was encouraged to engage more intimately with the science of risk.
This engagement, combined with their commitment to the public's right to
know about radiation, resulted in the production of novel risk representations that embodied the interests of the public in ways that previous strategies for visualization had not. Unlike mainstream media risk representations,
Safecast's visualizations were comprehensive in the type and amount of information about risk they presented and didactic in their mission to inform
users about the chaotic distribution of risk and the different kinds of radiation risks they faced.

The evidence and discussion of the differences in risk representation provided here have implications for rhetoric/communication scholars as well as institutions tasked with communicating to the public about risk. For rhetoric and communication scholars, it reveals that research in risk assessment and communication needs to be expanded beyond a critical perspective, which engages with grassroots rhetoric and communication primarily by identifying the failings of institutional communicators to fully or meaningfully engage with the public. With the blossoming of citizen science and other grassroots efforts to communicate about risk on the Internet, this chapter suggests there are now opportunities for descriptively assessing the goals of grassroots organizations by exploring their strategic choices for risk communication and argument.

For institutions involved with public risk communication, this assessment suggests that they can no longer expect to be the sole source of public representation of radiation risk and that they should consider descriptive assessments of grassroots communication, like this one, as opportunities to learn about the informational needs of the public. A review of the literature on radiation risk reporting from institutional sources in the aftermath of Fukushima suggests that their post hoc assessments of the accident include very little conversation about its communicative dimensions and no information at all about the efforts of citizen-science groups like Safecast to develop novel, citizen-centered risk communication strategies. In the few instances where institutional discussion of the accident does turn to issues of communication, the primary focus of discourse is on the success or failings of institutions or the mainstream media to effectively communicate about the accident. In the *Bulletin of the Atomic Scientists*, for example, media analyst Sharon Friedman comments on the general improvements made by the mainstream media in their reporting on the accident. She writes, "the Internet . . . gave the traditional media many opportunities for better coverage [of the Fukushima accident], with more spaces for articles and the ability to publish interactive graphics and videos" (Friedman 55). In a separate discussion in the same publication, Yoichi Funabashi and Kay Kitzawa[22] find fault with officials for not adequately communicating risk to the public. They argue, "The majority of the general population had no idea of the meaning behind the reported radiation levels. There was no yardstick against which to judge whether or not levels were dangerous. The government made no effective effort to educate or soothe the public in this regard" (Funabashi and Kitzawa 10). In none of the scholarly or official documents assessed[23] did analysts or representatives of the media, nuclear industry, or the Japanese government discuss the emergence of online representations of risk created by extrainstitutional groups and their implications for risk communication and messaging. As the case of Safecast suggests, the rise of these groups could represent a challenge and/or an opportunity for radiation risk communication. By ignoring the emergence of grassroots data-gathering organizations, institutional organizations, particularly governments, run the risk of facing communicative competition in a time of national crisis. By studying and engaging with groups like Safecast who are dedicated to developing representations for radiation risk, which incorporate the public need for comprehensive and didactic risk communication, institutional actors might learn how to improve their capacity to connect with the public in a time of crisis. By embracing the idea that public engagement with science can open up new lines of communication and critique that support public interests, scholars of rhetoric and communication might find themselves well positioned to facilitate these sorts of institutional-public collaborations in the dawning digital age of citizen science.

3

Information for and by the People

The Internet and the Rise of Citizen Expertise

As the previous chapter illustrated, the Internet allowed the members of RDTN/Safecast to have access to data and engage with scientific methods and information in a way that was previously unimaginable before the digital age. With the help of digital resources, they were able to create their own devices for gathering radiation data and design uniquely citizen-centered radiation risk representations. The transformative role of the Internet and Internet-connectable devices in RDTN/Safecast's capacity to develop citizen-centered risk communications raises questions about whether, and if so how, these technologies might also have influenced their ability to engage in public argument about radiation risk. In the last decade, the influence of the Internet on political argument and communication has increasingly drawn the attention of rhetoric and communication scholars who are interested in understanding its potentially transformative role in political participation and deliberation. In *Rhetoric Online: Frontiers in Political Communication* (2012), for example, Barbara Warnick and David Heinemann explore a range of subjects on this topic, including the use of new media in the 2008 election and the influence of viral YouTube video on the legislative debate over Don't Ask Don't Tell. In a similar fashion, Ian Bogost's *Persuasive Games* (2007) investigates the role of the Internet in political messaging by examining the use of online games to communicate about issues such as the American war in Iraq and the famine in Darfur. In addition to books on the topic, a number of articles have also explored the intersection of politics and the Internet covering everything from the use of the Internet to engage the electorate (Davisson 2011) to its role in fostering deliberative democracy (Ishikawa 2002).

Despite the growing interest in the Internet's influence on political deliberation and communication, there has been little effort to examine how it might be impacting public discourse and debate over issues with strong techno-scientific dimensions. If the Internet and digital media can provide more and better citizen engagement with science, then presumably it should also contribute in some cases to the development of more robust technical argument from the lay public. This chapter explores the question, How might changes in technologies for data gathering impact the capacity of laypeople to argue as experts? To do so it examines how the public arguments of the citizen-science group RDTN/Safecast were transformed by their participation in Internet-enabled citizen science. It makes the case that Safecast's program of Internet-enabled radiation measuring expanded its available means of ethical persuasion transforming its public arguments advancing and defending its citizen-science activities.

CITIZENS, SCIENCE, AND THE INTERNET

One of the fundamental changes the Internet has made in science in the last five years is that it has provided broader opportunities for nonscientists to participate in the production of scientific knowledge. From an office cubical or the comfort of a living room couch, anyone with a computer can contribute to the discovery of new protein configurations on the website FoldIt or help astronomers identify cosmic objects at Galaxy Zoo. As scientists increasingly turn to the public for help with exponentially growing piles of unprocessed data, there is hope that the public's involvement in scientific research will begin to break down the walls of expertise and authority that have separated science and scientists from the public. In a November 2011 article in the *Boston Globe*, Pulitzer Prize–winning journalist Gareth Cook strikes a note of optimism about the democratizing effect of the Internet in his piece "How Crowdsourcing Is Changing Science" where he writes, "Science is driven forward by discovery, and we appear to stand at the beginning of a democratization of discovery. An ordinary person can be the one who realizes that a long arm of a protein probably tucks itself just so; a woman who never went to college can provide the crucial transcription that reveals a spidery script to be a love poem from 2,000 years in the past" (Cook, "How Crowdsourcing").

Though hope is growing that the Internet and crowdsourcing will democratize science, the prospect of increased engagement has also raised the question, To what extent and in what ways will the relationships between the public and science be transformed through this new mode of interaction? In many cases the answer is "not much." Though scientists have invited the public to participate in research endeavors, they remain firmly in charge of setting research agendas and establishing the parameters and methods for fact gathering. Evidence of the unidirectional nature of this engagement is present in the discussions of researchers who famously used the work of FoldIt players to identify a candidate configuration for protease M-PMV PR (simian AIDS): "The critical role of the FoldIt players in the solution of the M-PMV PR structure shows the power of online games to channel human intuition . . . to solve challenging scientific problems. . . . The ingenuity of game players is a formidable force that, if properly directed, can be used to solve a wide range of scientific problems" (Khatib et al. 3).

Despite the enthusiasm for crowdsourcing science and the recognition here of the capacity of nonexperts to participate in scientific-knowledge making, support for their participation is tempered by doubts about their capacity to do science without guidance. As Khatib and coauthors note, game players are a formidable force in science only if they are "properly directed," suggesting that Internet crowd-sourced science may not be as democratic an undertaking as portrayed in media accounts.

Though many of the efforts to involve laypersons in scientific research are not breaking down the boundaries between the two, there are some rare cases in which the traditional relationships between science and the public does seem to be transformed by the technological affordances of the Internet. RDTN/Safecast represents just such a case. What separates it from most other citizen-science projects is that it was conceived and developed organically out of the individual and collective needs for information about risk. Further, members of the group made it their mission not just to educate themselves on the scientific literature about risk but to actually collect risk data and develop their own Internet-connective devices for measuring it.

The collective's rare status as a grassroots enterprise engaged in scientifically informed data collection and visualization presents a unique opportunity to explore the role of the Internet in transforming public discourse and argument on a techno-scientific issue. To examine this transformation, this chapter compares the *ethical* appeals used by Safecast and its predecessor RDNT in public discourse. Based on this comparison, it concludes that as the collective's Internet-enabled engagement with scientific information, methods, and tools for data collection increased, its repertoire for ethical argument was broadened from strictly nontechnical ethical appeals to both technical and nontechnical *ethos*. These newly available lines of argument were used by the group to publicly defend and advance its representations of radiation levels in the wake of the Fukushima accident.

Defining Expertise

Before examining how RDTN/Safecast's Internet-enabled citizen science broadened its repertoire for ethical argument, it is necessary to establish the essential characteristics of nontechnical and technical *ethos*. Because the difference between these categories of *ethos* is fundamentally a question of expertise, it is first necessary to ask and answer the question, What separates an expert from a nonexpert? Perhaps the most detailed investigation of this question is taken up by science and technology studies (STS) scholars Harry Collins and Robert Evans in their book *Rethinking Expertise*. In the text the authors offer a "periodic table of expertises" that lays out in detail the variety of dimensions and gradations of expertise which "individuals might draw on when they make technical judgments" (11). In the section of their table labeled "Specialist Expertises," the authors list a range of expertise that includes, sequentially, "beer mat knowledge, popular understanding, primary source knowledge, interactional expertise, and contributory expertise" (14).

The category "beer mat knowledge" represents the very lowest degree of expertness that can be obtained. At this level the information about a techno-scientific subject is at the quality of a factoid, fun fact, or brief explanation that might be found on a beer coaster or fast food placemat. This knowledge is considered largely unproductive by the authors, because it doesn't allow its possessor to do or make anything, debate the nature of the subject matter, correct mistakes made by others about it, or deliberate about the risks associated with it (Collins and Evans 19). The next position on the scale is occupied by "popular understanding," which "can be gained by gathering information about the scientific field from the mass media and popular books" (Collins and Evans 19). At this stage of knowledge, the user understands the meaning of a technical subject in a much more comprehensive way. As a consequence, they can explain it to others in a more sophisticated and less rote manner than someone who has just picked up a snippet of fact from a coaster. The next higher degree of specialization comes from engaging with "primary source knowledge." For Collins and Evans, the division between experts and laypersons emerges in this category, because both groups can engage with the specialist and semispecialist[1] literature of a particular technical field. While both groups are reading the same literature, the layperson's reading experience is not considered equivalent to that of the expert, because "if someone wants to gain something even approximating to a rough version of agreed scientific knowledge from published [primary] sources one has first to know what to read and what not to read; this requires contact with the expert community" (Collins and Evans 22). In other words, the reading experience of the expert and nonexpert are not equal because the expert reads as a *participant* in a particular culture of expertise and the nonexpert does not.

In the final two categories on the scale "interactional expertise" and "contributory expertise," the quality of this participation is what determines the level of specialization achieved by a subject. Persons judged to be at the level of "interactional expertise" have gained a specialized understanding about how scientific knowledge is made by observing scientists and their practices in a sustained and systematic fashion. Collins and Evans include in this category sociologists and anthropologists of science as well as journalists, sales persons, and managers whose work requires them to interact regularly with scientists (31–32). Based on their interactive experiences, experts at this level of specialization are capable of communicating clearly and in a nuanced fashion about the work of science in their area of expertise to experts in their own and other disciplines including other scientists (34–35). At the highest position on the scale, the category of "contributory expertise" requires not only immersion in scientific culture but also "the ability to *do* things within the domain of expertise" (24). It is these two qualities of expertise, interaction with the culture and contribution to the knowledge production of science, that will guide the theoretical distinction I intend to make between technical and nontechnical ethos.

Defining Technical and Nontechnical *Ethos*

In Aristotle's *Rhetoric*, he explains that *ethos* is an appeal to inspire confidence in the orator by highlighting his virtuousness (*arête*), goodwill for the audience (*eunoia*), or good judgment (*phronesis*) (I ii 1378a 6). To have a comprehensive account of technical *ethos*, it is important to consider what technical and nontechnical versions of each kind of ethical appeals might be. Collins and Evans' discussion of expertise perhaps most closely aligns with the category of *phronesis* or good judgment and, therefore, provides insights into the distinction between technical and nontechnical phronetic argument. As the authors explain, laypersons can always gain a higher degree of expertise by familiarizing themselves with the literature of a scientific field. However, without interacting in a sustained way with the culture of science or participating directly in the production of scientific knowledge, their expertise cannot rise above "primary source knowledge" and their *phronesis* or good judgment is always vulnerable to the criticism that they lack the full experience of the cultural context or practice of knowledge production to make prudent technical evaluations. Though laypersons might claim "interactional expertise" on the basis of their communications with acknowledged specialists, these interactions are typically temporary and consultative rather than sustained efforts to learn about the social, institutional, and epistemological culture of a field. As a consequence, these kinds of claims might be more appropriately characterized as nontechnical ethical appeals to *expertise by association* than "interactional expertise." Additionally, laypersons might claim to have "productive expertise" in a field because they have engaged in more disciplined observations or manipulations of nature as an *amateur, enthusiast, or hobbyist*. However, these kinds of claims can be considered nontechnical ethical appeals to *semi-expertise*, because participants can assume only a rudimentary understanding of the methods and procedures of a field, and their activities are commonly driven by personal interest rather than social or scientific exigencies. In contrast, arguers who have "productive expertise" can draw on their own *participation* in or *detailed understanding* of field-specific knowledge-making practices to construct their credibility. Their work is driven by a public interest in solving social or scientific problems or answering questions deemed relevant within a specialized field. And they can make a case for their good judgment using their *qualifications* such as their track record of publication, official title, awards, or honorific positions within a field of expertise.

Like arguments from *phronesis*, ethical appeals to *eunoia*, or goodwill toward the audience, can be made in a number of ways. Aristotle explains that in nonexpert argument an audience judges the degree of goodwill in a speaker on the basis of whether or not they believe he is being a friend to them. This belief is inspired if the audience feels that the speaker wishes the best for them for their own sake rather than for the sake of some benefit on the part of the speaker (II, iv 1381a 9). It is also inspired when the audience believes that they and the speaker consider the same things good and evil or that "they are like ourselves in character and occupation" (II, iv 1381b 15).

Whereas Aristotle understood goodwill as identification between the speaker and the audience, Roman rhetorical theorists discussed it in terms of a speaker's ability to condition the audience to be predisposed to accept his arguments. In his discussion of exordia in *De Inventione*, for example, Cicero explains that the beneficial conditioning of the audience can depend on *how* the orator presents his case. Audiences are more predisposed to accept the orator's case if he can present it in a manner that makes the audience attentive and receptive to his arguments. To make the audience attentive, Cicero counsels the orator to "show that the matters which we are about to discuss are important, novel, or incredible, or that they concern all humanity or those in the audience" (I, xvi, 23). To make them receptive, the orator must "explain the essence of the case briefly and in plain language" (I, xvi 23).

Goodwill of the kind Aristotle and the Roman rhetors discuss has a uniquely scientific or technical manifestation in the practice of *methodological transparency*. Though methodology would also be a dimension of *logos*, it can be considered a central feature in establishing goodwill for the speaker, because transparency about the procedures allows the audience to replicate or judge the process by which the expert arrived at their conclusions. Through methodological transparency, scientific or technical experts also establish identification with their expert audiences by illustrating that they are doing science according to the conventional practices of the knowledge community. Further, the explanation of methodology can also be judged as a gesture of goodwill toward the audience in the classical Roman sense of *benivolentia*, because the expert explains what might otherwise have been obscure about the process of knowledge creation. Conversely, any expert considered to have neglected, or worse yet, to have withheld information about their method would at the very least create suspicions about their membership in the knowledge community or, more seriously, face the charge of willfully obscuring erroneous or fraudulent processes. In either case, making an argument for one's conclusions or against the conclusions of one's opponents using *methodological transparency* requires special knowledge of procedures. In the case of laypersons, *eunoia* is limited to *methodological expediency* or negative attacks on the goodwill of scientists for making methodological choices based on inappropriate moral or pragmatic grounds. In Fabj and Sobnosky's investigations of AIDS, for example, activists' ethical arguments challenging the experimental protocols for AIDS drug testing rested primarily on the experiences of members of the AIDS community with clinical trials and on the moral and practical problems they associated with clinical methodology (176–77). Ethical charges of methodological obscurity or problems of identification are very seldom if ever raised by lay audiences, because they lack the experience or group membership to level these charges.

Like *phronesis* and *eunoia*, *arête* also has a unique kind of expert ethical appeal that is epitomized by the virtue of *epistemological objectivity*. Though there are a number of virtues that might also be considered expert or technical virtues such as precision, humility, originality, or skepticism,[2] none of these are as essential to the scientific identity as objectivity nor do they generate as many ethical attacks on or defenses of the virtue of technical expertise. Objectivity in technical *arête* is different from objectivity in the nontechnical ethical appeal to virtue. In nontechnical *arête*, objectivity is typically linked to the issue of the selfishness or selflessness of the speaker's motives as they relate to the common categories of material, social, or political desire. A nontechnical ethical attack on an expert's virtue, for example, might be made on the grounds that they received some sort of material benefit from their scientific work. This argument commonplace has been used regularly, for example, against climate-change scientists by nonexpert arguers who ethically critique these scientists on the grounds that they are generating results that keep them funded for further research rather than reveal anything meaningful about the influence of humans on climatic conditions. An ethical attack on objectivity by a technical arguer, however, would be *epistemological*, that is centered on the question of bias or the degree to which *a priori* beliefs or values about a social or natural phenomenon might influence the outcome of observation or experiment. Objectivity here is about disciplined knowledge-making rather than self-restraint in the face of opportunities for social, political, or material gain. Like *qualifications, participation,* and *methodological transparency*, the type of appeal to the virtue of *epistemological objectivity* is an instance of technical ethical appeal, because it requires the arguer to participate in or have intimate technical knowledge of expert procedures and epistemological practices.

The distinctions between technical and nontechnical ethical arguments from *phronesis*, *eunoia*, and *arête* provide a theoretical framework for judging changes in the status of ethical argumentation in the transition from RDTN to Safecast. The argument advanced here is that in the first phase of the organization's development as RDTN—in which they visualized data but did not participate in collecting it—their ethical argument was primarily nontechnical. Once they had begun to engage in the practice of collecting data in the second phase as Safecast, however, the nature of their ethical appeals changed and began to exhibit more distinctly technical characteristics. This transformation suggests that Safecast's participation in the Internet-enabled practice of data gathering opened up new lines of technical argument that had previously been unavailable to them as RDTN, and that the grassroots group seized on the opportunity to use these new lines of argument to defend and advance their representations of radiation levels in public discourse and argument.

Phase i: RDTN and the Centrality of Representation and Nontechnical Argument

As the previous chapter made clear, in the very early stages of its development the collective RDTN that was to become Safecast was a loose-knit confederation of individuals whose personal interests motivated them to collaborate on the development of a website for representing the radiation risks of the Fukushima accident. The website RDTN.org was created by Marcelino Alvarez and his colleagues at Uncorked Studios as a response to what Alvarez felt was a lack of essential information about radiation risk. In addition to Alvarez and his colleagues, the trio of Sean Bonner, Joi Ito, and Pieter Franken were also involved. They were personally invested in RDTN because they believed that it was a way of helping their friends and families inside Japan better understand and respond to their situation. The trio played an important role as facilitators and publicizers of the site. Ito and Franken provided Alvarez and his colleagues with information as well as connections to specialists that could help them with the technological development of the site. Bonner wrote about Safecast in the online media creating buzz about the group's activities.

As RDTN, the collective that would become Safecast was dedicated to the task of visualizing radiation levels across Japan. In this role they faced both misrepresentations of and attacks on their *ethos* in the mainstream media to which they had to respond. The primary targets of these misrepresentations and attacks were the "citizen scientists" who supplied the data on radiation levels for the website. In defending themselves, the collective relied heavily on nontechnical ethical appeals to *semi-expertise*[3] and *association to expertise*. The adoption of these defensive strategies suggests that though the group endeavored to identify themselves as a technical rather than a populist knowledge-making enterprise, they did not have the requisite credentials to defend this kind of representation.

It seems likely that the misrepresentations of and attacks on RDTN from the media were the consequence of the group's characterization of itself as a credible source of reliable information about radiation risk. From the launch of RDTN.org, Alvarez and his colleagues at Uncorked Studios were clear that the goal of the site was to provide reliable information about radiation risk to media consumers who were encountering a variety of incompatible reports about it from other sources. On the website's blog they explain, "With conflicting reports of radiation levels in affected areas, we wanted to build a way to report and see data in an unbiased format" (qtd. in White). The phrase "unbiased format" is a notable element in this statement, because it characterizes the website as an objective source of information in the techno-scientific sense of objectivity. However, the phrase "report and see data" that precedes it suggests that the meaning of "unbiased" here is complex. It does not mean that the data presented in the map are unbiased or objective in the scientific sense of the value, rather it intends that the website provides readers with a format that neutralizes the biases of the data in the same way that news reporters pursue objectivity in their stories: by presenting multiple perspectives on a single event. The online map makes this journalistic objectivity possible by visualizing and juxtaposing radiation data from a variety of sources both institutional and private. The idea is that if multiple sources agree on the level of radiation, then the radiation level must be correct. If they disagree, then the jury is still out on what the real level of radiation is. The creators of RDTN.org are very careful in the textual discourse on their website to dispel the impression that they are offering themselves as an authoritative source of data in contradistinction to existing institutional sources. Instead, their aim is simply to put forward all the data available so that visitors to the site can come to their own conclusions about radiation levels. As David Ewald of Uncorked Studios explains, "[Our data are] not, nor should they be considered a replacement for official information. This site supplements information by providing several data feeds" ("Open Dialogue").

Although RDTN.org's creators argued that their site is unbiased because it provided a diverse accounting of radiation levels, many of the news stories about the site gave the impression that RDTN.org was a reliable, unbiased, and genuine source of information, because the data on the site came from "the people" rather than from institutional sources such as the Tokyo Electric Power Company (TEPCO) or the Japanese government. Citizens were represented by these news sites as reliable sources of information in the nontechnical sense of objectivity, because they had nothing politically or materially to lose or gain by reporting radiation measures. In addition, they were con-

sidered to have the experiential authority to talk about the risks in their immediate environment. Here *phronesis* and *arête* have their sources not in a
subject's participation in disciplined scientific observation but rather in the
presence of the knower within a particular social, physical, or political context. These characterizations were in evidence, for example, in a web news
feed from *Time*, in which the author opened the article with the line, "Worried about the 'real' radiation levels around the damaged Fukushima plant?
Ask the people" (Travierso). Even though placing the word "real" in quota-
tions here suggests irony, when read in the context of the rest of the article
it is clear that this line is a sincere suggestion by the author that RDTN's
citizen-gathered information represents pure information, unpolluted by the
machinations of politics that cast a shadow of doubt and suspicion on the official, conflicted, and therefore unreal scientific measurements. In fact, immediately following the line the author writes, "There have been conflicting
reports on the level of actual radiation. . . . A new website [RDTN.org] aims at
filling the information gap" (Travierso). In these lines the phrase "conflicting
reports" references the discrepancies in radiation reporting from TEPCO and
the Japanese government that left a truth vacuum or gap where real information needed to be filled in. By posting the latest on-the-ground readings made
by "the people," presumably Japanese citizens, RDTN.org was characterized
as filling in these gaps with their measurements of the "real" radiation levels.

   In other articles, representations of the presumed trustworthiness or real-
ness of the data on the RDTN site were based on both the proximity of the reporters of the data to the radiation event and the timeliness with which they
could report it. In the first article published about RDTN, for example, Safecast cofounder Sean Bonner wrote, "The site allows people to submit their
own reads. . . . This way anyone can quickly get an idea of what is happening on the ground, first-hand" ("RDTN.org"). Aaron Saenz struck a similar
tone in his piece in the online news site *Singularity Hub* writing, "Several
crowd-sourced radiation maps have arisen that attempt to give up-to-the-
minute looks at the threat level in the areas most likely affected by the catas-
trophe. . . . This is a preview of how accelerating technologies will allow us
to monitor anything, anywhere, in real time" ("Japan's Nuclear Woes"). For
Saenz and Bonner, the timeliness and on-the-scene affordances of radiation
websites like RDTN add to their credibility because of the proximity of data
gatherers in time and space to the risk. Further, both authors suggest that
these affordances are a consequence of the populist nature of the site that
harnesses the power of the crowd to gather information.

   This brief overview of the early news reporting on RDTN.org and the organization's characterization of their mission suggests some dissonance between the way the group valued itself and positively constructed its credibility, and the way it was represented in the online media. In particular, media

sources seemed to attribute the credibility and objectivity of the site's data to its source—the lay public living in the radiation risk zone. In reality, of course, 80% of the information reported by the website came from the Japanese government. Of the remaining 20%, it is unclear how many sources were actually private citizens unaffiliated with local government or academic institutions.[4] On its website, RDTN.org made an effort to establish credibility for its enterprise by emphasizing the variety of its risk information sources rather than the timeliness or firsthand experience of the risk reporters. This fact raises the question, How could the values attributed to the site be so misconstrued? Despite the developers' efforts to explain the value of the site in terms of information diversity, this message was overshadowed by the prominence of the "crowdsourcing" frame in news reporting. This frame—which had become popular because of the Arab Spring that immediately preceded and was still unfolding at the time of the Fukushima accident—epitomized a narrative of citizen engagement facilitated by online social media. In this narrative citizens take power into their own hands through direct action against prevailing institutions and by documenting the events they generate through the Internet. The prevalence of this socio-political narrative is evidenced by the fact that the word "crowdsource" appeared in headlines associated with ten of the eleven stories about RDTN.org during this period.

It is also evoked through direct comparisons between Safecast and the news gathering efforts of citizens participating in the Arab Spring. NPR commentator Richard Knox, for example, explicitly makes this connection in his piece about Safecast: "Last we checked there were 100 or so data points [on RDTN.org] from citizen scientists, official Japanese sources, and a data aggregating site called Pachube. . . . There's no question crowdsourcing is a great way of finding out what's going on in real time by aggregating the input of lots of people in Tahrir Square, Tehran or Bengazi. But what about technical data. . . . Might that require some expertise of a kind citizen scientists shouldn't be assumed to have?" ("Citizen Scientists").

In these lines Knox makes an explicit connection between the crowdsourcing of news in reporting on the Arab Spring and the crowdsourcing of radiation measurements done by Safecast. He limits his comparison, however, by suggesting that though crowdsourcing is a valuable source of information about socio-political activities, it should not be considered reliable when applied to technical subjects like radiation levels. Knox's critique of the comparison between Safecast and the crowds documenting events of the Arab Spring reveals the extent to which the crowdsource frame informs his perspective on RDTN. Knox's concern over the reliability of the crowdsourced radiation data overlooks a fundamental fact that he himself asserts, which is that the crowd includes "official Japanese sources" as a primary constituency. Assuming that Knox is really not implying that official sources are tech-

nically unqualified to gather radiation data, his comment about the "expertise of citizen scientists" suggests that the "crowdsource" frame is guiding
his characterization of RDTN. The frame seems to encourage Knox to consider the RDTN crowd, like the Arab Spring crowd, to have *experiential* rather
than "contributory" expertise, or nontechnical credibility gained from being a
firsthand witness to an event rather than the technical credibility developed
through the disciplined production of scientific knowledge. The influence
of the crowdsourcing frame was not limited to Knox. In fact, it generated so
many similar inquires over the quality of the data presented on RDTN.org
that David Ewald of Uncorked Studios was forced to respond to the framing
issue directly. In the blog post "Open Dialogue," which addresses a number
of questions about the site, he writes, "We understand the sensitive nature
of this data and are approaching it as such. Crowdsourcing, while an un-
doubted buzzword in the media, implies that unfiltered data is being pulled
and displayed as fact" (Ewald).

The framing of RDTN as a "crowdsourcing" enterprise, while intended
as a complimentary (albeit mistaken) evaluation of the site's expertise and
virtue by most media pundits, invited critiques from others like Knox who
argued that the crowd could not be trusted to provide accurate information
about radiation risk. In Knox's report he challenges the validity of the website's risk representation by questioning the quality of the data based on the
reliability of its source. He comments, "Last we checked there were 100 or so
data points [on RDTN.org] from citizen scientists, official Japanese sources,
and a data aggregating site called Pachube. Is this a good idea? Or the latest
example of garbage in producing garbage out?" ("Citizen Scientists"). The
garbage in/garbage out question reveals that the primary problem Knox has
with RDTN is the quality of its data. At its core this problem of data is also
a critique of the credibility or *ethos* of the data gatherers. For example, Knox
uses the phrase "citizen scientist" to draw the distinction between credentialed
professionals and citizen dabblers in science in the statement "But what about
technical data. . . . Might that require some expertise of a kind citizen scientists shouldn't be assumed to have?" Here Knox places the phrase "citizen scientists" in contradistinction with "technical data," suggesting that the former
does not have sufficient expertise to produce the latter. In addition, his analogy
between citizen science and the citizen crowdsourcing of information in the
Arab Spring suggests that the authority of citizen scientists is tied up with
their experiential participation in a particular socio-political context rather
than with their role as dispassionate and disciplined observers of an event.

As the media circulated critiques of RDTN's *ethos* based on a "crowdsourcing" frame, the organization's creators pushed back. They offered their own
ethical representations of the site's contributors that attempted to recast them
as *semi-experts* who followed scientific procedures for ensuring the reliability

of their data and as persons with *associations to expertise*. These ethical defensive strategies appeared in the online media for the first time in interviews with Marcelino Alvarez on March 30 and April 1, 2011. In these interviews, Alvarez made an effort to defend the credibility of RDTN and its contributors by limiting the size of its pool of participants and reframing the enterprise as "citizen science" rather than "crowdsourcing." The former strategy debuted in *New Scientist* where technology reporter Jacob Aaron explains, following his discussion with Alvarez, that "taking radiation readings with Geiger counters is a fairly specialized activity, so RDTN's 'crowd' is smaller and more niche than many other collaborative websites" ("Japan's Crowdsourced"). With this comment, Alvarez, as reported by Aaron, suggests that the citizens providing the crowdsourced data for RDNT.org are specialized: limited to individuals that have the necessary experience in taking proper radiation measurements. Though the argument here is tautological—that the contributors to RDTN are doing a specialized activity, therefore, they are specialists—it nonetheless seeks a specialized or *semi-expert* status for RDTN contributors by differentiating them from typical crowdsourcing endeavors that have no specific criteria for crowd membership.

A more direct and detailed instance of Alvarez's efforts to defend the expertise of RDTN's participants appears two days later in a radio interview with *Far West FM*. In this interview, Alvarez provides an explicit definition of what he believes a citizen scientist is. This definition is meant to clearly uncouple the "crowdsourcing" frame from his organization's activities. The host Dave Brooksher begins by introducing the distinction between "crowdsourcing" and "citizen science" that Alvarez wants to make. He states, "RDTN is an upstart radiation monitoring network that allows civilians to find, in real-time, data about radiation levels around the world. . . . The site has already earned a reputation for *crowdsourcing*, but Alvarez finds that term somewhat inaccurate" ("RDTN.org Peer-reviews"). Immediately following Brooksher's introduction, Alvarez challenges the framing of the site as a crowdsourced endeavor by characterizing RDTN's contributors as citizen scientists: "We're specifically asking for people to contribute that might be more inclined to consider themselves as a 'citizen scientist' . . . which is an individual who either through professional or hobbyist means approaches the problem of capturing radiation in an academic way" ("RDTN.org Peer-reviews").

What is notable about Alvarez's definition here is that by defining RDTN's participants as "citizen scientist[s]" he is making a case for their *phronesis*, a move that stands in contradistinction to Knox's application of the term as a potential paradox or challenge to the notion that the members of the crowd could legitimately assume the title or expertise of a scientist. In Alvarez's definition the expertise of RDTN's citizen scientists has its source in their approach to data gathering, which is "academic." He also attempts to make the case for their "contributory expertise" by labeling some of them as "professionals." Despite these efforts, Alvarez is aware that there is a limit to how far he can push his characterizations. This is evidenced by the inclusion of the *semi-expert* label "hobbyist" in his description of his contributors. Also, later in the interview, he frankly admits that the site's contributors have no specialization in fields directly related to radiation: "We are not nuclear experts, and we are not health scientists." After making this disclosure, however, there is an effort to strengthen the authority of RDTN's citizen scientists by foregrounding their *association* with recognizable institutional experts and expert processes for vetting knowledge. Brooksher explains, "Data is checked before being published. They're [RDTN are] reaching out to members of the scientific and academic communities who can apply a peer-review model to the [radiation] readings" ("RDTN.org Peer-reviews").

In its initial phase of development, RDTN.org, which would become Safecast, was focused primarily on visualizing rather than gathering data about radiation in the aftermath of Fukushima. The developers of the website represented it as a democratic online space where radiation measurements from citizens and institutions alike could comingle to create a multisourced pool of information from which users could draw their own conclusions about the status of radiation risk in post-Fukushima Japan. Despite efforts by RDTN to make the case for the virtue of the site on the basis of its diversity and inclusiveness, the media offered an alternative characterization using the "crowdsourcing" frame that located the site's reliability in the quality of the crowd, which was represented as lacking political motive and having immediate spatial/temporal experience of risk. These characterizations supported the perspective that the crowd's data on risk was more reliable than institutional accounts of radiation levels; however, it also invited criticism that the site was sourcing its data from a nonexpert multitude. The creators of RDTN felt that they had to defend the credibility of their contributors using ethical appeals. By casting them as "professionals," they tried to distinguish contributors from the hoi polloi of other crowdsourced endeavors. In addition, they reinforced the expertise of their niche crowd by associating them with scientists and scientific practices for vetting data. These efforts to claim expertise, however, were tempered by the reality that RDTN was dedicated exclusively to data representation rather than data gathering. As a consequence, they could not make unqualified appeals to "contributory expertise" without losing all credibility. They, therefore, were forced to rely on nontechnical ethical appeals to *semi-expertise* and *association to expertise* to defend their *ethos*. With the transition of RDTN to Safecast and its move from data gathering to data collection, however, a new campaign of ethical argument emerged in which unqualified technical ethical appeals to "contributory expertise" and "interactional expertise" played a central role.

Phase 2: Safecast and the Development of a Sensor Network

As the previous section hints, the role of the Internet in transforming citizen argument is a complex one. Even though the Internet enabled RDTN to create and widely broadcast visualizations of radiation levels, something that had never been done by laypeople in the pre-Internet age, it still relied on strictly nontechnical ethical appeals to defend its ethos and the credibility of its risk representation. The expansion of the group's available means of persuasion from nontechnical to technical ethical appeals required more from the Internet and the group than simply the capacity to create and broadcast visualizations of radiation levels. It demanded that both the technology and its users be engaged in the scientifically informed practice of data gathering. As the previous chapter details, April 2011 was a time of transition for the confederation that made up RDTN. By the middle of the month, the group had added to its ranks Akiba and a few other members of Tokyo Hackerspace who had dedicated themselves to creating a radiation sensor network in Tokyo. By month's end, the enlarged group changed its name to Safecast and claimed as its primary mission both gathering and representing data about radiation risk. By taking the initiative to understand the methods of radiation measurement and to create new devices for measuring it, the participants in Safecast were able to open for themselves new lines of technical ethical argumentation.

Alpha, Beta, Gamma, and Eunoia

An assessment of media reporting on Safecast's data-gathering enterprise from its inception at the end of April 2011 to the end of December of the same year provides evidence that the collective's engagement with the methods and instruments of radiation measurement opened up new lines of technical ethical argument that the group used to critique institutional radiation risk reporting and advance the value of its own efforts. This assessment suggests that there were two areas in which their engagement with scientific practice and methods informed the group's representations of itself and the institutional sources it critiqued. These areas included the types of ionizing radiation that should be measured and the scale or scope on which measurements should be made.

In the transition from RDTN to Safecast, there is a distinct shift on the subject of radiation and its measurement from "primary source knowledge" to "contributory expertise" as well as new technical ethical appeals to *methodological transparency*. The shift from "primary source knowledge" to "contributory expertise" can be witnessed in the change in the discourse about radiation measurement. Early postings on RDTN's website about radiation are exclusively dedicated to rebroadcasting facts about radiation from scientific primary sources. A posting on April 7, 2011, for example, by Jacqueline Yanch, former professor of radiation and health physics at MIT, about the dangers of radiation is representative of the kind of discourse produced during this period. The post provides visitors to RDTN's website with a very accessible overview of the different kinds of ionizing radiation (alpha, beta, and gamma) and explains the dangers as well as the misconceptions about the dangers of these different radiation types ("Background Information"). Though the post offers evidence of the group's "primary source knowledge" about the different kinds of radiation as well as goodwill for the audience by offering an explanation of its nuances, there is no evidence of "contributory expertise"—a nuanced discussion drawing on the group's participation in or detailed understanding of field-specific knowledge-making practices.

In a blog post on May 5, 2011, however, Safecast's Sean Bonner signals a shift from "primary source knowledge" to "contributory expertise" when he introduces methods for measuring different types of radiation used by the group: "The sensors we are using are very sensitive and pick up alpha, beta, and gamma radiation which is different for some equipment and published readings which sometimes eliminate alpha altogether, or focus specifically on high energy gamma" ("Alpha, Beta, Gamma"). In this comment Bonner emphasizes the credibility of Safecast's radiation measurements by highlighting the sensitivity of their measuring equipment and the comprehensiveness of their evaluation. At the same time, he alludes to the problems with institutional measurement practices and risk representations by arguing that they either don't measure different kinds of radiation or they measure them but don't publish them. This response suggests a more-than-amateur knowledge of measuring practices as well as an effort by the group to participate in informed critical discourse about these practices. Similar critical discourse on measuring and reporting radiation appears in a number of instances where the organization makes the case for the credibility and importance of its own efforts. At a presentation on Safecast at the MIT-Knight Civic Media Conference on June 24, 2011, for example, founding member Joi Ito offers a pointed ethical attack on institutional *phronesis* by highlighting the Japanese government's disregard of variation in ionizing radiation. In a blog describing Ito's conference presentation, Ethan Zuckerman reports, "Most Geiger counters, Joi tells us, don't measure all three types of radiation—alpha, beta, gamma. They generally just measure gamma, which is the one most people care about. . . . But isotopes that throw off alpha and beta particles can be very dangerous when ingested. Japanese inspectors have taken to scanning bags of rice with gamma detectors and proudly announcing they're gamma free. That's irrelevant—the concern is that the food might have isotopes that give off alpha and beta particles. Joi suggests that the country is suffering from 'radiation illiteracy'" ("Mohammed Nanabhay").

Here Ito reportedly directly attacks the technical *phronesis* of official Japanese scanning methods that check food for only gamma radiation when the presence of alpha and beta radiation could be equally dangerous. In the context of his talk, this critique also sets up an argument for Safecast's expertise. Immediately before attacking the government, Ito offers a brief description of the kinds of sensors Safecast is using and the efforts they have made to measure separately all three kinds of radiation ("Rock in One Hand").

These technical ethical constructions of Safecast's *phronesis* also include a second, closely related line of ethical argument from *eunoia*. This argument is an appeal to *methodological transparency,* which functions in two ways to support the efforts of Safecast and critique the Japanese government. First, by explaining their measurement methods Safecast shows goodwill toward the reader by demystifying a technical process, which plays a significant role in risk evaluation. Second, through their explanation they also reveal that institutional sources have failed to educate the public and their representatives properly about the risks of radiation resulting in "radiological illiteracy." This lack of education is not the consequence of a failure of the expertise on the part of the government: they know that there are three different kinds of radiation that could be measured. Rather, it is a failure of goodwill: they neglect to train their representatives in the proper methods of measurement, and, in doing so, obscure the risks of radiation from the public. This technical ethical appeal to methodological transparency contrasts with the group's earlier engagement with methodology as RDTN. In this earlier phase, there was no substantive ethical argument based on a critique of the government's methodology or praise of their own methods. The fact that Safecast seizes on *methodological transparency* to critique the ethos of the government here suggests that it recognizes the value of these technical appeals in making the case for its citizen-science endeavor and is able to cross the divide between nontechnical and technical argument to employ them.

While Safecast's ethical appeals evidence a transition from "primary source knowledge" to "contributory expertise," this "contributory expertise" does not have precisely the same character as the one Collins and Evans associate with scientists. For them, scientific "contributory expertise" typically "enables those who have acquired it to *contribute* to the domain to which the expertise pertains" (*Rethinking Expertise* 24). Specifically, it enables them to advance scientific knowledge in that domain by testing hypotheses or answering research questions. If this is the case, then how can Safecast and its members—who have experience in data gathering and knowledge about measurement practices but who are not themselves pursuing a scientific research agenda—be accommodated in their model? Cases like Safecast's could find a place in their typology if the concept of "contribution" was divided into "analytical" and "technical/informational" contributions. The subcategory "technical/

informational contributory expertise" would include the efforts of groups like Safecast who have acquired expert-level procedures and methods and gained experience in data gathering but whose contributions are guided primarily by social rather than scientific exigencies. Their goal in gaining expertise is to contribute to the solution of public problems that arise from the techno-scientific risk and require them to understand scientific information, methods, and technologies. These characteristics distinguish them, on the one hand, from *amateurs* or *enthusiasts* who might gather data with less technical knowledge and discipline and who pursue private rather than social or scientific interests. On the other hand, they would also differentiate them from "analytical experts" who belong to an expert community for whom broader social exigencies are secondary to or in parallel with the development of new scientific knowledge and practices.

Safecast's history and its representation in the mainstream media suggest that "technical/informational contributory expertise" is a more reliable descriptor for their activities than "analytical contributory expertise." The influential role of social exigence in Safecast's mission is evidenced in the history of the group's formation. As the discussion in the previous chapter explains, the members of Safecast became experts on radiation because the people they loved were threatened by radiological risk. As their sensor network grew, their pragmatic motivations also expanded from helping their friends and family understand their radiation risk to helping all Japanese citizens. It was only after they had begun to accumulate substantial data to address the social exigencies of radiological risk that they started to consider their activities as potentially contributory to science. Even then however, they characterized them as informational rather than analytical. This representation comes through in reporting on the group's activities. In the NBC News story "Japan's Citizen Scientists Map Radiation, DIY-style," for example, journalist Amanda Leitsinger writes about Safecast's goals: "They have assembled thousands of radiation readings plotted on maps that they hope one day will be an invaluable resource for researchers studying the impact of the meltdown" ("Japan's Citizen"). Here Leitsinger characterizes Safecast as an *informational resource* for scientists studying the accident rather than as a group of researchers with a scientific agenda. This representation is reinforced later in the article by Safecast volunteer Brett Waterman who explains, "In 10 or 20 years' time, you can't go back to three months after the event [at Fukushima] and then find out what the data was like. But if you record it now, and then we continue to record it over the months and years to come, then from a scientific and community point of view there is a database that can be referenced" (qtd. in Leitsinger, "Japan's Citizen").

In these lines, Waterman characterizes Safecast's goal as the creation of a database of radiation measurements that scientists and the community might use to draw conclusions about radiation risk. There is no hint in this explanation that the group is gathering the data as part of an independent analytical agenda. By emphasizing data-gathering activities and implying that these activities are sufficiently disciplined to generate useful data for scientists, both Waterman and Leitsinger highlight Safecast's "contributory expertise." However, their representations suggest that Safecast's contribution falls into the category of "technical/informational contributory expertise," because the group's data-gathering activities, though disciplined, are not connected to a research agenda or a particular scholarly conversation within the scientific community.

The Institutional Response to Safecast

With the increased intensity of Safecast's data-gathering activities and their technical critique of the Japanese government, it is natural to imagine that there would be efforts by the government to defend itself and launch counterattacks on the organization. In the only direct government response to the efforts of Safecast, MEXT representative Naoaki Akasaka characterizes the government's perspective on the group's work in the following way: "We think it is beneficial to get much information from many sources about radioactive contamination level[s] at their living region, but specific information sources are never recommended by us" (qtd. in Leitsinger, "Japanese Government"). In framing the benefit of Safecast's work in terms of the quantitative argument "more is better than less," the government both embraces the group's efforts while at the same time avoids sanctioning the quality of their data. This response appears unusually restrained given the vitriolic charges made by the group of the government's "radiological illiteracy"; however, given the socio-political context in which the Japanese government was asked to comment on the group's efforts, their measured response appears to be a well-considered rhetorical move.

There are a number of reasons the Japanese government might not have decided to publicly attack or attempt to discredit Safecast. One of the reasons could be the government's own problematic track record of reporting radiation levels after the Fukushima accident. For example, it failed to release or use predictions produced by SPEEDI—a system of sensors measuring radiation across Japan that detailed the path of the fallout in the days immediately following the accident—as a guide for planning initial evacuations. As a consequence, villages north of the plant directly in the path of the radiation plume were considered safe and were used as evacuation sites (Onishi and Fackler). In addition to putting citizens in harm's way, the government also drew public ire for not disclosing evidence that the core of the reactors had melted until months after it was known and for changing the level of radiation permissible on school grounds from one to twenty millisieverts so that they could limit the relocation of student populations (Aoki). By early June, roughly a month prior to Leitsinger's interview with MEXT, public opinion polls were indicating that 80% of Japanese respondents did not trust the government's communications about the crisis (Krieger). In this context it seems reasonable that the Japanese government would want to avoid making ethical attacks on other groups doing radiation monitoring, particularly grassroots groups like Safecast that had a populist appeal. It may also account in some part for why Safecast representatives adopted a more strident offensive stance toward the government's ethos rather than simply defending the technical credibility of their own project.

Though Japanese government officials were restrained in their comments on and critiques of Safecast's risk assessment, a less directly focused but more vigorous attack on citizen efforts to measure radiation was published by the *Japan Times*. The story "Experts: Leave Radiation Checks to Us: Laypersons just Spread Fear with Inaccurate Readings, They Say" defends the credibility of the government's radiation data-gathering efforts and attacks laypersons for taking their own radiation measurements. Central to the attack on lay radiation measurers is a contrast between their measuring devices and methods of measurement and the government's. According to Genichiro Wakabayashi a professor of radiology at Kinki University who was interviewed for the article, lay measures often rely on "cheap and easy-to-handle devices sold on the Internet [that] can sometimes show abnormally high radiation levels" (qtd. in Matsutani). These types of devices are contrasted with the "large, expensive and high-quality equipment" used by the Japanese government that "determine radiation levels more precisely than small, cheap devices" (Matsutani).

This challenge to the validity of lay radiation measurements predicated on the quality of the equipment is a dialectical counterpoint to the defensive arguments made by Safecast about the credibility of its own measurement practices. In fact, Safecast's choice of ethical appeal to "technical contributory expertise" was a direct response to the arguments in Matsutani's article. Conversations on Safecast's Google discussion board reveal that members of the group read this article and responded to it. Whereas most items on the board commonly elicited one or two responses the discussion of this particular article drew fifteen comments ("Experts"). Though the majority of the responses revolved around the general question of Safecast's ethos and how to preserve it, some of the comments responded directly to the article's criticism that the high radiation levels measured by these groups were a consequence of the low quality of their equipment. In these responses the primary line of defense was that perhaps the readings of lay measurers were high because their monitoring devices and methods were more sophisticated than the government's, not less. This dialectical counterattack appears, for example, in a response by Akiba from Tokyo Hackerspace who writes, "You have people like the author of the article worrying that the people reading higher than normal numbers on their Geiger counters will cause a panic. On the other hand, he should ask himself why those numbers are higher than regular dose numbers from the government. If high energy gamma were the only thing to worry about, then the numbers should not be different at all" ("Re: [Safecast Jpn]"). With the statement "if high energy gamma were the only thing to worry about," Akiba suggests that one of the major problems of the government's measurement practices is that they neglect to take into account low-level alpha and beta radiation. This oversight could be as much a consequence of the methods for measuring radiation as it could be the type of equipment being used. With this reasoning, Akiba turns the critique—that lay radiation measurers obtain elevated risk measurements because of the low quality of their devices—back on the government as a criticism of the quality of their measuring devices and practices. The appearance of this line of argument here as well as in the discourse of other members of Safecast suggests that by participating in data gathering they could and frequently did draw on technical ethical appeals to "technical contributory expertise" and *methodological transparency* to defend their radiation-gathering efforts as well as attack those of the Japanese government.

Arête and Ad Numerum: *Who Needs Credentials When You've Got Axioms?*

In addition to critiquing the quality of the measuring equipment used by laypersons, institutional representatives also challenged the expertise and character of the risk reporters by attacking their long-term commitment to radiation measurement. These qualitative attacks threatened to undermine reliability of the data they had gathered. To answer these critiques, members of the group turned to technical ethical appeals to *arête*.

Institutional critiques of the commitment and reliability of noninstitutional measurers and Safecast's responses to these critiques appear in mainstream media sources. In the *Japanese Times*, for example, Professor Genichiro Wakabayashi of Kinki University challenges the commitment of noninstitutional measurers. He argues, "The important thing is to keep monitoring at the same place over a long period of time to check changes in radiation levels. Thus, the figures from the Education, Sports, Science and Technology Ministry [MEXT] are after all reliable" (qtd. in Matsutani). In contradistinction, he characterizes the measurements made by laypersons as being few and fleeting wrapped up in the moment of crisis rather than in a long-term commitment to gathering radiation data:

> Magazines and Internet content, including personal blogs, articles and postings of monitoring results by individuals sometimes slam the science ministry for publishing results deemed meaningless.

> But Wakabayashi said monitoring over short periods is meaningless. The science ministry measures radiation levels every day. (qtd. in Matsutani)

The contrast made in these comments—between the fickle short-lived efforts of lay radiation measurers and the persistent measurement of the government in fixed locations over long periods of time—was a perspective to which Safecast felt they had to respond. In their response they argued publicly that though the government was providing continuous readings of radiation levels at a single spot in a town or city it was not reporting the radiation levels in detail at multiple locations across these populated areas. As a consequence, the group could make the argument that it was collecting data the government was neglecting and that the data it was collecting was more important to residents than the kind the government was gathering. Both of these appeals are colorfully made by Safecast's Sean Bonner in an interview with *Al Jazeera* on August 10, 2011:

> Sean Bonner . . . told Al Jazeera that [Safecast] volunteers have so far logged more than 500,000 radiation data points across Japan.
>
> He said his group was the only organization he knows that is tracking radiation on a local level. . . .
>
> "We spoke with a woman in Japan on Saturday who said since March she's been calling local offices, and the federal government, just trying to get data, and she has not been able to get a single reading close to her house," Bonner Said. "Part of that is that the information is just not there, the government doesn't have it." (Jamail)

Bonner's argument here seems to be a strategic response to the institutional critique that citizen measurers cannot be reliable sources of expertise about radiation risk because of their sporadic and fleeting attention to it and its measurement. The dialectical quality of this response is illuminated by considering it in the frame of Perelman and Olbrechts-Tyteca's two primary *loci*, or commonplaces of argument: the *loci of quality* and the *loci of quantity*. In the *New Rhetoric*, the authors define *loci of quantity* as "those *loci communes* which affirm that one thing is better than another for quantitative reasons" (85). *Loci of quality*, by contrast, "occur in argumentation when the strength in numbers is challenged" (89). The *loci of quality*, therefore, are defined by their status as foils to quantitative argument. To attack or defend against an argument from the *loci of quantity*, an arguer can invoke qualities like the essential, unique, precarious, or irreparable character of an object, action, state, or person. The *qualitative locus* of *precariousness*, for example, might be used to oppose the *quantitative locus of duration* on the grounds that "anything that is threatened acquires great value: *Carpe diem*" (91). The qualitative argument based on precariousness, therefore, stands in opposition to the basic quantitative claim "more of the same is better."

The argument advanced by the institutional sources in the *Japan Times*, however, is not suggesting that the precarious measurements of laypersons are superior because of their uniqueness. In fact, just the opposite is the case. They are claiming that that which endures is unique and superior to that which is numerous, fleeting, and trendy. This strategy of employing a *qualitative locus of duration* rather than a *quantitative* one is recognized by the authors of the *New Rhetoric*: "Besides the uses of the *locus* relating to the unique as original and rare, with precarious existence and the loss of which is irremediable . . . the *locus* of the unique is used in an entirely different connection, in opposition to the diverse. Here the unique is that which can serve as the norm and the latter takes on qualitative value compared with the quantitative multiplicity of the diverse. . . . [For example,] ancient authors offer fixed, recognizable models, which are eternal, and universal. Modern authors . . . have the disadvantage that they cannot serve as the norm, as a model beyond dispute: The multiplicity of modern authors makes them pedagogically inferior" (93). In this discussion of the *qualitative locus* of *duration*, Perelman and Olbrechts-Tyteca explain that the longevity of a particular phenomenon is a consequence of some essential universal quality that emerges from the cacophony of collective opinion on a subject after many years of trial and challenge. In other words, that which endures after many years of challenge and testing is said to be qualitatively superior to proliferate, untested, and trendy opinions on a topic.

Perelman and Olbrechts-Tyteca's complementary *loci of quality* and *quantity*, in particular the *qualitative locus of duration*, provides an argument scheme for identifying the dialectical relationship between the institutional critique voiced in the *Japan Times* and Safecast's ethical defense. In the *Japan Times* article, lay representations of risk were criticized for being not authoritative, because they were taken by a novel diverse crowd of measurers for a short period of time, whereas the representatives of the government had been measuring radiation regularly for many years. In order to respond to this ethical critique based on the *qualitative locus of duration*, Safecast turned to *ad numerum*, a quantitative appeal to the number of radiation readings their volunteers had done. Though uncommon, *ad numerum* has been used as a synonym for *ad populum* (appeal to the people), which is defined as an argument asserting that a proposition must either be accepted because of the number of people who believe in it or rejected because it is supported by the crowd or mob[5] (Walton 62). In the case of Safecast, I propose that a distinction be made between *ad populum* and *ad numerum*. For the purposes of this analysis, I will use *ad numerum* to refer to an appeal in which the arguer attempts to certify the validity of a claim on the basis of the number of data points collected by a crowd rather than the number or quality of the persons collecting the data. Shifting assessments of credibility from people and their quantity or quality to the numerousness of the data they collect involves switching from an *ethical* locus of subjective credibility to an objective or rational one. In the case of *ad numerum*, the objective rationale for credibility is described by Jakob Bernoulli's axiomatic central limit theorem, which holds that the calculated *a posteriori* probability of an event (p) gets closer to the true *a priori* probability of an event (P) the greater the number of trials (n) that are conducted. In other words, in the case of an argument *ad numerum* the basic principle of Bernoulli's limit theorem—that more data yields a closer proximity to the truth—is used to certify the reliability of the data (Daston 238).

The importance of *ad numerum* to the defense of Safecast's *ethos* is evidenced by the frequency with which it appears in discussions about the group's data-gathering activities in online media. Of the fifteen articles sampled after the middle of April when the group began to move into data gathering, this line of argument is made on at least four occasions.[6] Before the middle of April, the appeal is never used, suggesting that it is closely linked with Safecast's transition into data collection activities. In most cases the appeal *ad numerum* functions as a defense against challenges to the reliability of Safecast's risk measurements, a controversy directly connected with the technical expertise of the data gatherers. For example, *ad numerum* is used in an interview on the website *O'Reilly Radar* as a response to contributing writer Alex Howard's statement, "Crowdsourcing radiation data on Japan *does* raise legitimate questions about data quality and reporting, as Safecast's own project leads acknowledge" ("Citizen Science"). In response to this critique Marcelino Alvarez explains that the quality of the data is certified by its quantity: "We make it clear to everyone on our site that yes, there could definitely be inaccuracies in crowd-sourced data. . . . And yes, there could be contamination of a particular Geiger counter so the readings could be off. . . . But our hope is that with more centers and with more data being reported that those points that are outliers can be eliminated" (Howard). This same argument is taken up a few lines later and amplified by Safecast's Sean Bonner who argues, "More data is always better than less data. . . . Data from several sources is more reliable than from one source, by default. Without commenting on the reliability of any specific source, all the other sources help to improve the overall data" (Howard).

What Alvarez and Bonner emphasize in their statements is that the size of the data set will ultimately certify its validity erasing problems that might arise from the deficiencies of "technical contributory expertise" in the data-gatherers and irregularities in their instruments and methods of measurement. By credentializing their efforts *ad numerum*, the representatives of Safecast seem to be skirting the issue of *qualifications* altogether using a deductive axiom to sever the validity of the measurements from the credibility of the measurer. In the *Realm of Rhetoric*, Perelman discusses the possibility of such a strategy in his explanation of how the relationship between a speaker and his acts might be neutralized. He explains, "When a person uses a *method* to prove a truth, to establish a fact in an incontestable way, the *character of the person* who affirms the fact does not in any way modify the status of the information" [emphasis mine] (Perelman 96). That "method" for Perelman could include axiomatic reasoning is supported by his and Olbrechts-Tyteca's elaborated commentary on the speaker-act relationship in the *New Rhetoric* where they write, "Irrespective of his wishes and whether or not he himself uses connections of the act-as-person type, a speaker runs the risk that the hearer will regard him as intimately connected with his speech. This inter-action between speaker and speech is perhaps the most characteristic part of argumentation as opposed to demonstration. In formal deduction, the role of the speaker is reduced to a minimum" (317).

In the case of Safecast, the appeal to *ad numerum* seems to be an effort to short circuit any discussion about the credentials of gatherers by claim-ing that their status as nonexperts should have little bearing on the validity of their data. Instead, the data should be able to stand for itself credential-ized by its numerousness according to the axiomatic principle of the central limit theorem. In this way, the members of Safecast are appealing to tech-nical *arête*, in particular the virtue of objectivity, as a way of cutting short ethical challenges to the qualitative or subjective character of their data gath-erers. The perception that the group is attempting to sever the relationship between the quality of the data and the data gatherers on these grounds is captured in media characterizations of the group. Miles O'Brien of PBS, for example, illustrates this when he remarks, "Safecast believes people should trust in the data, and the more people who are gathering it the better" ("Safe-cast Draws"). Here the notion of "trust," which is central to credibility, is lo-cated in the data rather than the people gathering it. Further, we see a direct challenge to the *qualitative locus of endurance*, based on the quantity and di-versity of the popular masses, in the argument that "the more people who are gathering it the better": a clear invocation of the *quantitative more and the less* commonplace.

Safecast's use of *ad numerum* appeals to skirt the ethical issue of technical expertise offers important evidence that it has transitioned away from tradi-tional nontechnical ethical appeals to virtue that are considered the hallmarks of lay arguers in risk debate and deliberation and toward the argument style of institutional actors who frequently draw on objectivity as a defense against ethical challenges. This strategy of using an *ethos of objectivity* has been iden-tified by both rhetorical and historical scholars as a prototypical feature of institutional risk argumentation. In her insightful paper "The Presumption of Expertise: The Role of Ethos in Risk Analysis," Carolyn Miller, for example, examines how the Atomic Energy Commission (AEC) tried to use probabilistic risk analysis as a way of circumventing problems with its own *ethos*. The AEC hoped that by offering the American public a probabilistic risk assessment of the dangers of nuclear power they could maintain the credibility of their perspective on the safety of the technology despite the crisis of trust plaguing the US government in the wake of the Vietnam War and Watergate scandal (Miller 189–91). Similarly, historian Theodore Porter argues in *Trust in Numbers* that statistics began to replace judgments of character in policy arguments with the increasing size of bureaucracies and the consequent loss of intimate connections amongst bureaucrats (ix–xi).

Though quantification and mathematization have been considered havens for governments and government officials in times of ethical crisis, to my knowledge it has never been considered that grassroots organizations might use this tactic as well. I would maintain here that though Safecast is not making a technical ethical argument for "contributory expertise," they are in fact drawing on expert knowledge, in this case mathematics, to stand-in for their credibility. Because this practice is common for bureaucratic arguers but not commonly associated with public argumentation, Safecast is in a way mimicking the style of institutional argumentation where authority can be established through an *ethos of objectivity*. This appeal to virtue could not have been made by Safecast if it were not involved in a sustained effort, supported by the Internet and Internet-connectable devices, to collect data about radiation. It is reasonable to conclude then that the appeal to *ad numerum* is attributable to the group's development of "participatory expertise," which, though not directly generative of an ethical appeal, expanded their available lines of argument such that they could appeal to the technical *arête* of objectivity to work around the ethical problem of *qualifications* plaguing their data gatherers.

Conclusion

The development of the Internet and Internet-supported devices for gathering, manipulating, and representing data has given rise to an explosion of citizen-science activity that seems to be bringing scientists, scientific practices, and the public in closer contact with one another. Though this new proximity has not yet resulted in the democratization of science or the assumption of technical expertise by the demos, it has in some instances changed the nature of the relationship between citizens and science in public debate. By examining the case of Safecast, in which the democratization of science seems to have been realized to a greater degree than in most other cases, this chapter has endeavored to show that citizens who become involved in collecting data also begin to expand their opportunities for discovering and using technical lines of argumentation. In the case of Safecast, this expansion is evidenced by the transition in the group's argument strategies from predominantly nontechnical ethical appeals from *semi-expertise* and *expertise by association* to technical ethical appeals from *methodological transparency*, "contributory expertise," and *ad numerum*. This expansion suggests that perhaps the previously theorized obstacles to lay public argument—the lack of participation in scientific culture and practice—are being eroded thanks to the affordances of the Internet. Though this chapter examines in detail only a single case of this erosion, and is, therefore, limited in the scope of its conclusions, it encourages further investigation not only into the influence of the Internet on techno-scientific debates but also into the possibility afforded by the Internet for improving public deliberation on techno-scientific issues.

4

Warming Relations?

The Benefits and Challenges of Promoting Understanding and Identification with Citizen Science

Because the members of Safecast had expertise in designing visualizations and building sensing devices, the group provided an ideal case for examining how Internet-enabled citizen science could support the development of citizen-centered risk communication and improve the capacity of laypersons to participate in argument about techno-scientific issues. Although these outcomes are important aspects of digital-age citizen science, there are others that deserve attention. One area that invites investigation is whether citizen science can improve relations between laypersons and scientists. This chapter explores this subject by inquiring, Can citizen science be an effective means of promoting identification and mutual understanding between laypersons and scientists? and Can citizen science provide access for laypersons to technical sphere argumentation? Because the case of Safecast generally involves a novel engagement of citizens with *science* rather than citizens with *scientists*, this chapter turns to a new case study in which a citizen and a scientist work collaboratively to investigate an issue related to one of the most politically sensitive topics of the last quarter century: climate change. The endeavor, known as the Surface Stations project, was conceived of and developed by Anthony Watts, a critic of climate change, and Roger Pielke Sr., a climate scientist, to investigate siting conditions at temperature measurement stations across the United States. By exploring the discourse and argument generated by this citizen-science endeavor, this chapter suggests that citizen science can provide opportunities for laypersons to participate in the technical sphere. However, this access does not necessarily establish common ground or goodwill between scientists and laypersons, because entrenched socio-political values and epistemological commitments on either side introduce obstacles to the process.

Citizen Science and Its Outcomes: A Review of the Literature

The first of the two primary questions advanced in this chapter, Can citizen science be an effective means of promoting identification and mutual under-

standing between laypersons and scientists?, is not unique to this investigation. Research into whether, and if so how, participation in citizen science
might affect the relationship between laypersons, science, and scientists has
been conducted in a number of fields most notably Public Participation in
Scientific Research (PPSR) and Public Engagement with Science (PES). The
field of PPSR includes scientists and science education specialists interested
in studying how the public's involvement in citizen science might help improve their understanding of science and their capacity to identify with scientists. A general outline of the field's interests and research agenda appears
in *Public Participation in Scientific Research: Defining the Field and Assessing Its
Potential for Science Education* (2009). In this document the authors explain
that the field is rooted in the Public Understanding of Science (PUS) movement of the 1950s, which "was premised on the idea that science, scientists,
and other experts know and should determine what the public needs to learn"
(Bonney et al., *Public Participation* 10). They clarify, however, that PPSR differs in important ways from what has come to be pejoratively labeled the
"deficit model" of public-science engagement because it embraces the conclusions: 1) that people have greater motivation to learn about science if it is
relevant to their lives and if the learning process is interactive, and 2) that
focusing on content is not as important as focusing on the process of scientific research (10).

   As part of their scholarly investigation of whether, and if so how, citizen
science achieves these goals, PPSR scholars have developed a taxonomy of citizen-science projects, which they use to discuss the strengths, weaknesses,
and challenges of different models of citizen-scientist engagement. In their
assessment, PPSR scholars divide citizen science into three separate types:
contributory, collaborative, and cocreated projects. Of the three, *contributory
projects* have the lowest level of interaction between the lay public and the scientific processes involved. This designation contrasts with Collins and Evans'
"contributory expertise," because instead of developing expertise through activities generated from their own interests, lay participants' research agendas
and plans for action are set by scientists. In addition, though activities in the
categories of contributory projects and "technical contributory expertise" are
limited to data collection, projects designated "technical contributory expertise" can include some technical or procedural innovation by citizen scientists as we saw in the case of Safecast. The next category, *collaborative projects*, include public volunteers in the analysis of data and the dissemination
of results. In some cases, lay public participants even provide feedback on
experimental design. The final category of citizen-science projects, *cocreated
projects*, is considered by PPSR scholars to be the most interactive. In these
types of projects, the public are involved at all levels of the scientific process
from crafting research questions to designing methodologies for data gathering to disseminating conclusions and taking action. Projects at this level

contribute to the development of "contributory expertise" in laypeople; how-ever, they still differ from the citizen science described in the previous chap-ter, because they involve joint scientific and lay exigencies rather than exclu-sively lay exigencies.

With the help of these taxonomic categories, PPSR researchers have made some assessments of how effectively different levels of engagement influence the public's understanding of science. However, these efforts are relatively new and underdeveloped. In a rare if not unique comprehensive assessment of how well citizen-science projects achieve educational and relational goals, the authors of *Public Participation in Scientific Research* explain that they "quickly discovered . . . that few PPSR projects had been evaluated comprehensively, especially those . . . categorized as Collaborative or Co-Created" (Bonney et al., *Public Participation* 20). In their own assessment of the extent to which current citizen-science projects were reaching their goals for education and identification, the authors conclude that there was a direct correlation be-tween the type of project and the degree to which these goals were achieved. They argue that participants in contributory projects made the least head-way in achieving citizen engagement objectives, because these types of proj-ects involved members of the lay public already interested in science, and because activities in these projects were restricted to learning about scien-tific content rather than the processes of scientific investigation (46). Con-versely, the most important gains in citizen engagement came in collabora-tive and cocreated projects: "Our case studies suggest that projects which involve members of the public in the largest number of steps or categories of research, those that we have categorized as Collaborative or Co-Creative, yield the greatest impacts in terms of understanding science process, devel-oping skills of scientific investigation, and changing participants' behavior toward science and/or the environment" (48).

Though PPSR research identifies types of citizen-science engagement and provides a cursory overview of the educational/behavioral outcomes of citizen-science projects, conspicuously missing from their research are questions about the role citizen science might play in improving public debate and in shaping science and scientific attitudes. While PPSR scholars include "citizen action" and "community involvement" in their criteria for identifying differ-ent kinds of citizen science, they avoid discussing these political dimensions in detail (22). Authors of the *Public Participation in Scientific Research* report write, for example, "In our attempt to develop models of PPSR we have delib-eratively excluded . . . activities that involve members of the public in under-standing and influencing public policy as opposed to participating directly in research" (Bonney et al., *Public Participation* 19). They do, however, mention that citizens can contribute to argument in the technical sphere, though this contribution is limited to the quality of the data they can supply to scientists for their research: "Thousands of individuals have submitted data on nest-ing birds, the distribution and abundance of monarch butterflies, [etc.]. . . . These data have been of sufficient quality to allow scientific analyses and to be published in peer reviewed scientific publications" (45).

Whereas PPSR limits itself to the examination of the educational/relational benefits of citizen science for the lay public, Public Engagement with Sci-ence (PES)[1] embraces an approach to citizen science that includes its socio-political dimensions. In the report *Many Experts, Many Audiences: Public En-gagement with Science and Informal Science Education* (2009), PES scholars identify their perspective on the study of the relationship between science, scientists, and the public as a strong rejection of the Public Understanding of Science model of public engagement. In particular, they argue that whereas PPSR scholars see public engagement with science as a downstream pro-cess, where laypersons gain knowledge about science and its practices from scientists, PES scholars view this relationship as bidirectional. Further, they see more effective public deliberation and decision making as the end goal of this engagement: "Public engagement with science (PES) is usually pre-sented as a 'dialogue' or 'participation' model in which publics and scientists both benefit from listening to and learning from one another. . . . The model is premised on the assumption that both publics and scientists have exper-tise, valuable perspectives, and knowledge to contribute to the development of science and its application in society" (McCallie et al. 23).

Although PES scholars do not reject the value of citizen science as a way for scientists to engage laypersons, they argue that this engagement needs to move beyond education and the modification of public attitudes toward sci-ence to embrace public participation in social action and debate on techno-scientific issues. They maintain that research should focus more heavily on "the valuable knowledge and perspectives publics bring from their lives that enhance the discussions . . . of science related societal issues" than on the "need for publics to learn science" (McCallie et al. 24). Because of this shift in research focus, PES explorations of citizen science tend to concentrate on scientific or lay investigations informed by scientific practices that lead to civic actions and are guided by the interests and concerns of the lay public. Though there are currently few investigations of citizen science by PES scholars,[2] ex-isting case studies exemplify the differences in research agendas between PES and PPSR. Gwen Ottinger's article "Buckets of Resistance: Standards and the Effectiveness of Citizen Science," for example, examines the use of air sample buckets by community activists in Louisiana to bring attention to the problem of air quality in a community abutting a Shell chemical refinery. Ottinger de-fines citizen science as citizen-directed rather than scientist-directed research: "Bucket monitoring is typical of 'citizen science'—knowledge production by, and for, nonscientists—in that it both contributed information about local air quality and suggested alternative modes of air quality assessment" (245). In the case of the bucket activists, their efforts to collect and analyze air quality using scientifically sanctioned tools and methods represent a sort of guerilla science carried out to raise awareness of the shortcomings in existing scientific measurement protocols and to advocate for alternative practices. Though ultimately the efforts of these citizen scientists did not change scientific or bureaucratic standards or practices, Ottinger argues that the citizen-science project had value, because it "challenged the standard practices used by regulators for assessing air quality . . . [asserting] that peak toxics concentrations mattered to determining whether industrial emissions threatened community health" (245). Following this same vision of citizen science as a way of challenging institutional practices and conclusions, other PES scholars in the field of environmental justice have conducted investigations of the influence of citizen science in a variety of contexts including pesticide regulation, PCB exposure, and alternative energy advocacy (Ottinger and Cohen).

By investigating the influence of citizen science on political engagement, PES researchers seem to be covering the unexamined aspects of the relationship between citizens, science, and scientists not covered by PPSR researchers. However, their exclusive focus on *citizen-directed* citizen science misses the possible importance of collaboratively developed citizen-science projects, like the one undertaken by Watts and Pielke, on public debate. Further, because of their exclusive interest in public sphere arguments PES scholars, like PPSR researchers, leave unexamined the influence of citizen science on arguments in the technical sphere. The examination of the argument and discourse of the Surface Stations project, which follows, addresses these holes in the current scholarship by investigating a collaborative citizen-science endeavor between a layperson and a scientist that addresses an important sociopolitical issue: climate change. In addition, it explores with equal interest argument and discourse in the public and technical sphere generated by citizen science. By investigating these dimensions, it adds more detail to the current spectrum of research exploring the possible influence of citizen science on the relationship between citizens, science, and scientists. Finally, because it examines *how* citizen science and its results influence discourse and argument, it adds a new rhetorical methodological perspective to the existing repertoire of sociological and scientific methods of exploring these questions.

The Odd Couple: Pielke and Watts and the Surface Stations Project

PPSR researchers studying citizen science have concluded, at least tentatively, that the more collaborative a project is the more likely it will positively affect lay perspectives on and understandings of science and scientists. If we accept this finding, as well as the possibility that closer collaborations might open up the technical sphere to lay perspectives, it makes sense to begin this investigation by examining the character of the collaboration between the climate change critic Anthony Watts and the climate scientist Roger Pielke Sr. It seems hard to believe that such a collaboration could even exist, let alone involve a high level of mutual engagement. However, this is exactly what happened in the Surface Stations project. To understand how this collaborative citizen-science endeavor developed, it is necessary to examine in detail the kind of problem the two parties were trying to solve, the nature of their individual interests in the problem and the incentives for them to work together toward a solution.

The techno-scientific problem that the Surface Stations project was developed to tackle is related to a long-standing scientific quest to identify possible biases in surface temperature measurements. To calculate the average surface temperature across the United States, the National Oceanographic and Atmospheric Association (NOAA) relies on 1,221 weather stations distributed across the country called the United States Historical Climatology Network (USHCN) (Menne, Williams, and Vose 994). Because the data from these stations make up part of a broader data set used to assess and model climate change, the accuracy of these measurements is extremely important. This fact has not been lost on NOAA, which has taken pains to ferret out and statistically correct biases in the system introduced by conditions such as changes in measuring equipment and the times at which measurements are made. However, not all biases have been comprehensively accounted for.

One of the more neglected and potentially significant sources of bias is the influence that physical conditions at individual weather stations might have on temperature measurement. The lack of knowledge about site conditions and their possible influence on temperature measurement was recognized almost one-and-a-half decades ago by the National Academy of the Sciences (NAS). In the NAS report *Adequacy of Climate Observing Systems* (1999), authors for the academy explain that each observing station and its operating features needed to be documented including "station location, exposure, [and] environmental conditions" so that any biasing factors might be accounted for (17). The lack of information about local site conditions mentioned in the report was recognized and acted upon to some extent by NOAA, which cited the report as a motivating factor in its plans to develop a network of pristinely sited weather stations across the United States known as the Climate Reference Network (CRN), which would avoid significant measurement biases introduced by local environmental factors (NOAA, *United States* 10). In addition, researchers at NASA and NOAA made an effort to remotely assess the effects of siting related to urbanization (urban heat island effect [UHI]) and changes in land use and land cover (LULC). For one such effort, Hansen and coauthors (2001), for example, used satellite data of nighttime light emissions to determine the possible effect of urban encroachment on measurement sites. The authors of the research assumed that stations sited where light emissions were highest had experienced the most urban encroachment, and where they were lowest the least. Based on their analysis of the satellite data, they concluded that the bias of urbanization on temperature data was modest (approximately .1°C per 100 years), but encouraged further research into the topic (Hansen et al. 6).

Although the scientific institutions tasked with collecting and studying climatic data (NOAA, NASA, and NCDC) were aware that siting conditions might influence temperature measurements and took some action to investigate these conditions, their efforts were in no way comprehensive. There was, for example, no systematic effort to visit temperature monitoring stations and document their environmental conditions firsthand. Without on-the-ground observation, a totally accurate description of site conditions or their possible biases on temperature measurements could not be made. This lack of information, though not a priority for the programmatic research agenda of NOAA and NASA, still drew the attention of a handful of climate researchers. One of those researchers was Roger Pielke Sr., a climatologist at Colorado State. In 2001 Pielke, in collaboration with a number of other climate scientists mostly based in Colorado, began investigating whether and to what extent general climate models accurately reflected local climatic conditions. Toward this end, members of the research project visited eleven long-term weather stations in eastern Colorado and assessed how their local site conditions and history might have influenced the quality of the data from these stations. In their investigation, they found errors in the details of official site characterizations and noted that the current qualitative information about site conditions, or metadata, was sparse and sparsely used in regional climate modeling. They argued, as a consequence, that modelers of local climate needed to either clearly state the limitations of their model's validity or expand their efforts to include more qualitative information about the sites used in their models (Pielke et al., "Problems" 432). Their findings were published in a 2002 article "Problems in Evaluating Regional and Local Trends in Temperature" in the *International Journal of Climatology*.

In a follow-up paper on the subject in the *Bulletin of the American Meteorological Society*, Pielke and his graduate student Christopher Davey expanded their examination of weather stations across eastern Colorado from eleven to fifty-seven stations to broaden their understanding of site conditions. To judge site quality, the authors used exposure standards from the World Meteorological Organization (WMO) (1996), which recommended that temperature measurement instruments should be 1.5 meters above the ground and that sites "should be level . . . and should offer free exposure to both sunshine and wind (not too close to trees, buildings or other obstructions)" (Davey and Pielke 497). To document the conditions at each of the sites, Davey and Pielke "took at least five photographs. One was a picture of the temperature sensor itself. The other four illustrated the views from the temperature sensor in each of the four cardinal directions" (499). In addition to documenting site conditions, the authors examined a database of forms that included existing metadata about the site.[3] Their investigation revealed that "sites that met all the WMO site exposure requirements were in the minority" (498). Because problematic siting conditions were so pervasive, the authors questioned whether they might be introducing systematic biases into the long-term temperature data set. As a consequence, they recommended that 1) NOAA consider removing poorly sited stations from the Historical Climatology Network, 2) climatological and meteorological researchers "determine whether there is a systematic warm or cold bias from site exposure," and 3) their technique of photo documentation "be extended to the entire USHCN network as well as to all surface stations worldwide used in long term temperature trend analysis" (503).

By visiting sites and documenting their conditions, Pielke, Davey, and the other contributing researchers had uncovered what appeared to be a potentially serious systemic problem in the USHCN and the data it produced. Because the data from the network were the bases for constructing local, national, and international climate models, Davey and Pielke's conclusions could impact the credibility of government institutions responsible for these measurements as well as the conclusions about climate change they supported. Given the gravity of their findings, it is not surprising that representatives of the National Climatic Data Center (NCDC)—the federal center in charge of gathering, storing, and assessing climate data—responded. What was surprising, perhaps, was the antagonistic tone they adopted in their response. In the same issue of the *Bulletin* where Davey and Pielke's research appeared, representatives of the NCDC admitted that the site conditions in the Colorado survey were unacceptable and warranted further investigation: "We agree that some USHCN stations in eastern Colorado (and probably elsewhere) have inappropriate exposure for monitoring climate. . . . The USHCN database could definitely benefit from improved site exposure documentation" (Vose et al. 504). However, they offered very few details about what, if anything, the NCDC would do to rectify the problem. Instead, they challenged the idea that the data gathered by Davey, Pielke, and other researchers put into question the current climate models and methods of data analysis by arguing that 1) without further statistical investigation no connection could be made between site quality and biases in the data, and 2) the sample of stations in Davey and Pielke's study was too small and geographically limited to paint a comprehensive picture of the whole Historical Climatology Net-

work: "They [Davey and Pielke] did not quantify the temperature bias produced by the exposure problems, nor did they show that those problems actually resulted in spurious temperature trends. Furthermore their analysis
was a static assessment of site exposure over a relatively small part of the
country. . . . In other words their results do not show that a large number
of USHCN stations have a comparable exposure problem" (Vose et al. 505).

The challenges to Davey and Pielke's work raised by Vose and his coauthors explain, in part, Pielke's interest in collaborating to develop the Surface Stations citizen-science project. Pielke believed that there was likely a
systemic problem of poor siting conditions across the USHCN and that this
might affect temperature trend assessments. However, there was no way for
him to prove this without examining the conditions at a larger portion of the
1,221 sites in the network. Also, without significant amounts of data it would
be difficult to do a reliable statistical analysis to check whether, and if so how,
site conditions influenced temperature trends. Undaunted by the criticism
and heavy burden of proof, Pielke continued to research and write about the
station siting issue producing two more papers on the subject in 2007.[4] Although neither of these papers significantly moved beyond the work he had
already done, one of the papers, "Unresolved Issues with the Assessment of
Multidecadal Global Land Surface Temperature Change," was the seed from
which the Surface Stations project grew, expanding Pielke's research of siting
conditions and temperature measurement beyond even his own expectations.

On May 4, 2007, Pielke posted a description of his "Unresolved Issues"
paper along with a link to the original text on a climate science weblog[5] he
maintained as part of the scholarly outreach for his climate research group at
the University of Colorado, Boulder.[6] On May 7 the blog post just happened
to be read by Anthony Watts, a regular patron of the blogosphere who had
professional and personal interests in temperature measurement and climate
change. Professionally, Watts had some expertise in the science and technology of weather measurement and forecasting having been a TV weather
meteorologist for twenty-five years ("Anthony Watts"). He had also started his
own company ITWorks, which specialized in products for television and Internet weather and broadcasting (*ITWorks*). In addition, or perhaps as a consequence of his professional experiences, Watts also had a personal interest in
the issue of climate change, which he explored in his blog Wattsupwiththat.
com. In his posts Watts adopted a critical perspective on institutional climate
science and its conclusions that the earth was artificially warming because
of human activity. In a blog posted just months before his encounter with
Pielke's research, he wrote, for example: "[NOAA] released their report on
weather records of 2006 today. . . . In that report they state 'The 2006 annual temperature for the contiguous U.S. was the warmest on record' . . . I
don't doubt that a bit, nor do I dispute it. . . . But (and here it comes) *I don't*

think it has anything to do with man-made greenhouse gases" (Watts, "2006 Hottest Year").

Because of Watts's professional background in meteorology and his personal interest in climate change, he both understood and enthusiastically received the contents of Pielke's 2007 paper. On May 7, 2007, Watts posted a comment on Pielke's blog[7] explaining that he had read his paper "with great interest and much head nodding" ("Re: A New Paper"). In his comment he reveals that he too had been concerned with the influence of conditions at weather stations on temperature measurement and offered for Pielke's consideration other issues: "Poor siting and biased siting is adequately covered in this paper [Pielke 2007 "Unresolved Issues"], but there are many more examples your readers may not be aware of" (Watts, "Re: A New Paper"). One of these issues was the effect that the change from whitewash to latex paint might have had on temperature measurement. He explains in his comments that he had encountered the issue as an undergraduate at Purdue where as a student he had once questioned a professor about the apparently anachronistic use of whitewash on weather screens. The question earned him a lecture on the heat reflective properties of whitewash. After raising the question of paint with Pielke, Watts inquired about whether any research had been done on the subject. In a comment attached to his blog, Pielke replied that it had not and encouraged Watts "to publish a note" in the *Bulletin of American Meteorology* or other publications. In response Watts wrote, "What really needs to be done is a simple exposure experiment. . . . I might very well do that here and report some quick preliminary results. Then [I'll undertake] a study of some shelters that have been around for a long time and . . . [have] layers of paint on them" (Watts, "Re: A New Paper"). This exchange between Watts and Pielke about paint suggests a good interpersonal rapport between the two with Pielke offering encouragement for Watts's concerns and Watts taking action as a result of the scientist's advice. This initial positive dynamic set the stage for a more significant discussion about collaborating to document site conditions at weather stations in the United States Historical Climatology Network.

Encouraged by Watts's decision to study paint at instrument shelters, Pielke asked him to consider helping him with his research by cataloguing other details about the conditions at the weather stations he visited. He wrote, "If you visit sites [to inspect for paint], please also photograph them using the method we reported in our paper [Davey and Pielke 2005]" (Pielke, "Re: A New Paper"). Watts responded enthusiastically to Pielke's request by not only agreeing to visit sites but also volunteering to get other people involved in the project: "I'll get as many as I can in California and I may be able to put out a nationwide call to weather casters using our connections to the many TV clients our company [ITWorks] serves" (Watts, "Re: A New Paper"). In the days following this initial contact, Watts made good on his promise by documenting site conditions at instrument shelters using the methods described in Davey and Pielke's 2005 paper. Just two days after his initial communication with Pielke, Watts wrote his first blog entry describing his documentation of the conditions of the USHCN weather station at the California State University, Chico. He begins his description by explaining, "I visited the Chico State University Farm on Hagan Lane . . . to do a site survey in the format done by Dr. Roger Pielke of Colorado State University" (Watts, "Site Survey"). After identifying his method, Watts describes the conditions at the site including basic data, like exact geographic location and elevation of the site as well as complex qualitative information about the station's physical environment, such as its location close to an asphalt road and the presence of electronic equipment inside the station. Finally, following Davey and Pielke's method, Watts visually documents the weather station with a photograph of it from the north, south, east, and west.

Although neither Pielke nor Watts were fully aware of the magnitude of the enterprise that was about to unfold, the beginning of May 2007 was the moment of conception of a new citizen-science endeavor that was to be known as the Surface Stations project. The goal of this project would be to document the site conditions at every station in the USHCN network. To achieve this goal, the website surfacestations.org was created, which gave instructions for how to document sites and post images and data of siting conditions. The initiation of this larger site documentation project developed out of a sense of shared interest in how site conditions might bias temperature measurement and was helped along by the participants' recognition that by collaborating each of them could overcome significant obstacles to pursuing this interest. By collaborating with Watts, for example, Pielke could surmount his greatest research obstacle, the lack of investigative resources. Without substantial support from the NCDC, NOAA, or NASA, all of whom had shown only a limited willingness to invest time and resources to investigate the on-the-ground conditions at USHCN stations—Pielke couldn't marshal the necessary people-power to visit and document the conditions at each of the 1,221 sites in the network. However, because of Watts's business and professional connections as well as his presence in the blogosphere, he was in a position to raise the human resources needed to document the conditions at all of the USHCN sites.

In addition to having connections, Watts also had the technical know-how to set up an online site for communicating about the project and collecting observations. On May 17, 2007,[8] he began work on the website surfacestations. org. By June 4, 2007, the site went live with a call for volunteers "to photograph, survey, and catalogue every USHCN station for the purpose of doing a qualitative analysis on the near surface temperature data produced by the USHCN data set" (Watts, "Surfacestations.org Is Ready"). Within three months, over two hundred volunteers had signed up to help.[9] Though not all of these volunteers made observations, there were still enough active participants for Watts to have received data on 331 stations, or a little more than one fourth of all USHCN weather stations, within three months of the project's launch. By August of the following year, he had collected data on half the stations. At last count,[10] 1,068 of the 1,221 stations (approximately 87%) have been surveyed and 1007 rated (Watts, *Surfacestations.org*).

Though Watts was able to marshal the citizen scientists required to investigate the conditions at the USHCN weather stations, the fruit of their efforts would not yield a harvest of scientifically useful data without a rigorous system for documenting those conditions. To ensure that it could stand up to scientific scrutiny and participate in scientific argument, disciplined protocols for making and checking observations had to be in place. Pielke's primary contribution to the project was in helping design these protocols. Evidence of the climate scientist's participation in this aspect of the project appears in an article in the online edition of *Reno News & Review*. In the piece "Watts, Me Worry?" author Evan Tuchinsky reports that according to Watts, "The site [surfacestations.org] includes a set of standards for inspectors, for which he consulted [Roger] Pielke [Sr.]" ("Watts"). On the surfacestations.org website, Watts provides specific protocols for how to do a weather-station site survey. These protocols are the same ones used in Davey and Pielke (2005). They include recording the GPS location and elevation of the site as well as the height of the measurement instrument. They also involve taking photographs to "show things within a few hundred feet" of the weather station as well as pictures of the station from the four compass points. Additionally, observers are requested to measure "the distance from the shelter [i.e., weather monitoring station] to influences if close by, 25 feet or less" (Watts, "How to" 2). The term "influences" here means trees or man-made structures that might interrupt flat terrain or "inappropriately placed" things, such as air conditioners, parking lots, or bodies of water that might introduce bias into temperature measurement.

In addition to designing protocols for site documentation, Pielke also served as a resource for identifying criteria for ranking station quality based on their site characteristics.[11] The criteria used for station ranking by the Surface Stations project were develop by Michel Leroy[12] of Meteo-France (France's premier meteorological institution) but adopted by NOAA as part of its push to create a Climate Reference Network (CRN).[13] In NOAA's *Climate Reference Network Site Information Handbook* (2002), the organization lays out a scale for rating sites with "1" being the best and "5" the worst. The primary criterion for ranking stations "good" (types 1 and 2) or "poor" (types 3, 4, and 5) is the distance of their measurement sensors from artificial heating sources.

Whereas well-sited stations have no artificial heating sources (parking lots, air conditioners, bodies of water, etc.) within thirty meters, poorly sited stations reside anywhere from ten meters to right next to such sources (NOAA *Climate* 6). By including the distance of measurement devices from artificial heating sources as part of the site survey, Watts was able to apply NOAA's criteria to the data from volunteer station surveys to rate the quality of the USHCN stations. These ratings, as we will see, are pivotal in both the public and technical sphere argument about the quality of stations in the USHCN and their influence on temperature measurement.

By examining the history of the Surface Stations project and its development, it is possible to understand both the problem that the project was meant to address and the circumstances that brought together the climate scientist and climate change critic to investigate it. The project emerged organically as a consequence of mutual interest and aid with both the lay person and scientist contributing resources essential for advancing toward a shared goal. Because of the mutual interest and involvement of both parties, this project could be reasonably ranked as a *cocreated* citizen-science project on the scale developed by PPSR scholars. Given the character of the collaboration and the conclusions drawn about these kinds of collaborations in PPSR scholarship, we would expect that this project would improve identification between Watts and the climate science and climate scientist with which he was working. We might also expect that by working with a climate scientist Watts might be able to participate in the technical sphere debate over siting bias and temperature measurement. The rhetorical examinations that follow of the discourse and argument in the public and technical spheres generated from this citizen-science collaboration suggest that the project's outcomes in these areas are mixed and that the variability of the consequences might be attributed to the influence of socio-political as well as institutional values and goals on discourse and argument.

A CITIZEN'S ACCOUNT: ANTHONY WATTS'S
IS THE U.S. SURFACE TEMPERATURE RECORD RELIABLE?

As the Surface Stations project got underway and accreted a critical mass of site documentations, Watts, Pielke, and other scientific researchers began to generate discourse and argument in the public and technical sphere based on the results of the project. The next few sections examine the most remarked upon text that developed from the project: Anthony Watts's report *Is the U.S. Surface Temperature Record Reliable?* (2009). In the report, Watts tells the story of the Surface Stations project's inception, describes the data generated from its investigations of USHCN sites, and draws conclusions about climate change from the evidence. Close examination of the report's discourse and argument reveals that though Watts and Pielke worked together to conceive and build the Surface Stations project, their citizen-science collaboration did not inspire Watts to identify with or even recognize Pielke's extensive contributions in his report. The analysis that follows suggests that Watts's substantial dissociation of Pielke and his contributions from the Surface Stations project is encouraged by Watts's personal critical perspective on climate change and the views of his conservative audience about the credibility of institutional efforts to measure long-term temperature trends.

Government Neglect and the Citizen-Science Hero

In order to make the case that Watts's values and the values of his conservative audience might have influenced his representation of Pielke and his contribution to the citizen-science project in *Is the U.S. Surface Temperature Record Reliable?* it is necessary first to identify exactly what those values are. Watts's personal commitment to a critical perspective on climate change and his exigence for involving himself in the Surface Stations project has already been established in the previous background section. However, the question, How might a conservative audience construe the value of the citizen-science project and its relationship to the climate change debate? remains unanswered. Media coverage of the project in the conservative press preceding the publication of Watts's report provides an important resource for answering this query. In the period between the beginning of the project and the publication of the report, the Surface Stations project was written about in online and print media a total of ten times.[14] An assessment of the sources of this coverage reveals that the majority of them—eight out of a total of ten articles—appeared in the conservative press. A more detailed analysis of the contents of the articles about the Surface Stations project appearing in the conservative press reveals that two major themes pervade their reporting: the valorization of the citizen scientist and the insinuation of government neglect. In the first media report on the project,[15] for example, conservative columnist Bill Steigerwald of the *Pittsburgh Tribune-Review* characterizes Watt's project as a response to the government's neglect to properly investigate the weather stations it uses to collecting climatological data. He writes, "Anthony Watts . . . suspects NOAA temperature readings are not all they are cracked up to be. . . . He has set out to do what big-time arm-chair climate modelers like Hansen and no one else has ever done—physically quality-check each weather station to see if it's being operated properly" ("Helping"). In his account, Steigerwald is correct when he explains that no one has physically checked *all* the weather stations. However, it is notable that he never mentions that some stations had been checked directly by climate scientists, like Roger Pielke Sr., and remotely by NOAA and NASA years before the Surface Stations project. Instead, Steigerwald portrays Watts's work as a novel solo underdog effort meant to uncover the corruption of temperature data which the government had neglected to address.

After the story of the Surface Stations project broke in Steigerwald's article, Watts and the Surface Stations project got a lot more attention in the conservative press. In the days following the initial reporting on the project, the story was picked up by Fox News and the Drudge Report, which increased the traffic to Watts's website to twenty thousand hits in a single day. The stories in the conservative press that followed also invoked the theme of neglect and added valorizations of the citizen scientist as well. In an article on the Heartland Institute's website on November 1, 2007, for example, James Taylor, the organization's columnist on environmental policy, wrote, "Anthony Watts . . . believes in sound science. So much so, in fact that, he singlehandedly created a volunteer army of citizen-scientists to make sure climate scientists are receiving the most accurate information available regarding U.S. temperature readings. Unfortunately, the scientists who compute the nation's average annual temperature seem to have little interest in obtaining accurate information" ("Meteorologist"). In Taylor's reporting here, both the neglect and citizen-scientist hero frames are used to describe the Surface Stations project. The term *singlehandedly* is evocative of the heroism of the undertaking as well as the use of the phrase "volunteer army," which figures Watts as a military hero taking troops into battle. The battle itself is a fight over ignorance that climate scientists have conspired to propagate by neglecting the effect of local conditions on historical temperature measurements. Thus, the heroism of Watts, and his citizen-science army, is linked with government neglect.

Whereas the conservative media valorized Watts and his populist army of citizen scientists and suggested a government cover-up, in the more centrist/liberal media, alternative themes of citizen-scientist cooperation and collaboration emerged. Unlike in the conservative media, Pielke's role in the Surface Stations project is preserved in the reporting. The first article in which Watts and Pielke are mentioned in the same news piece appears in the centrist[16] *Oroville Mercury-Register* at the end of June 2007. In the piece, Pielke's previous work on investigating siting conditions and Watts's collaboration with the scientist are described by staff writer Ryan Olson. Based on information from an interview with Pielke, Olson credits the climate scientist with pioneering the research that inspired Watts's work, "He [Roger Pielke Sr.] said Watts's work is serving a need to know how the stations gather data. Pielke's previous research has shown many weather stations have been poorly placed" ("Watts' up?"). In addition to recognizing Pielke's earlier work, Olson also highlights the collaborative relationship between citizen and scientist when he writes, "Pielke's research group maintains its own blog. He and Watts have corresponded and posted entries about each other's efforts" ("Watts' up?").

A similar recognition of Pielke's role in developing the project appears in the left-of-center *Reno News & Review*. In Evan Tuchinsky's interview with Watts, the reporter notes that Pielke supplied Watts with a method for site evaluation. In addition, he reinforces Pielke's priority in discovering the problem as well as the collaborative nature of his relationship with Watts. He explains, "He [Watts] wasn't—and isn't—alone in his scrutiny of official weather stations. Professor Roger Pielke Sr. . . . also was looking into U.S. Historical Climatology Network. . . . His research group was posting his findings online, and his efforts dovetailed nicely with Watts's" (Tuchinsky).

The media coverage of the Surface Stations project in both the conservative and centrist/liberal press before the publication of Watts's *Is the U.S. Surface Temperature Record Reliable?* reveals that there are clearly two different ways in which the citizen-science endeavor is being represented. In the conservative media, the narratives of the citizen-scientist hero and government neglect, dominate, and elide the efforts of Roger Pielke Sr. and NOAA and NASA to investigate the problems with the siting of temperature measurement devices. In the centrist and liberal media, however, the collaborative nature of the Surface Stations project is recognized by including Pielke in their coverage. The difference in reporting in the liberal/centrist and conservative media provides us with a set of value frames representing views of the conservative audience about the significance of the Surface Stations project.

Now that the conservative and centrist/liberal framings of the citizen-science project have been established, it is necessary to examine in detail the discourse and argument in Watts's *Is the U.S. Surface Temperature Record Reliable?* to establish whether and, if so, in what ways these framings may or may not have taken hold in the text and what their participation reveals about the capacity of Watts's report to publicly promote understanding and identification between laypersons and scientists. Perhaps the most striking set of rhetorical choices illustrating the influence of conservative values on Watts's representation of the Surface Stations project appear in the introduction of the report. In the opening sections "Whitewash versus Latex" and the "Story of Three Stations," Watts tells the tale of how the problem of poor siting conditions was discovered as a prelude to his description of the Surface Stations project and its findings. A close textual analysis of the choices made in these initial sections reveals that Watts's narrative of the discovery of the problems at USHCN stations substantially diverges from the historical record of the event. Most notably it includes no mention of his communication with Dr. Roger Pielke Sr., the climate scientist's efforts to explore the problem of station siting, or the efforts of any government institutions to investigate the problem.

The erasures of Pielke and institutional science from the problem discovery narrative in Watts's report are evidenced in the first line of the first section

"Whitewash versus Latex." Opening the section Watts writes, "The research project described in this report was the result of pure serendipity. It began when I set out to study the effect of paint changes on the thermometer shelters, known as Stevenson Screens" (4). What is notable about Watts's opening line is that it attributes the discovery of the problem partially to chance and partially to his own inquiries into the effects of paint on temperature measurement. As the research narrative in the section unfolds, there is no further indication that Watts's knowledge of the problem came from any other source but his own experience. In fact, an analysis of the grammatical subjects in all of the sentences in the first two sections reveals that the first person singular pronoun "I" appears a total of twenty-four times and the possessive pronoun "my" a total of ten times. In contrast, the pronouns "we" and "our" in reference to any collaborative work with Pielke appear zero times.[17]

By removing any mention of Pielke and his efforts to bring awareness to the potential problems of station conditions on temperature measurement, Watts creates a problem discovery narrative in which his own efforts as a citizen scientist play a critical role. This revised narrative suggests that through a bit of serendipity and common sense Watts, as citizen-science sleuth, was able to discover problems that government scientists had either neglected or ignored. In the second section, "A Story of Three Stations," Watts elevates himself and the figure of the citizen scientist by attributing the discovery of problematic siting conditions to his investigation of the effect of paint on temperature measurement. He explains, "Next [following the paint experiments], I set out to determine if the Stevenson Screens of the U.S. network of temperature monitoring stations had been updated to latex paint" (5). With this line, Watts leads the reader to believe that his motivation for investigating site conditions was driven internally by his natural scientific curiosity about the effects of paint on temperature measurement. The communications between Watts and Pielke clearly show, however, that though Watts did visit the USHCN stations because of his interest in the effects of paint choice on screen temperature, it was Pielke who had made him aware of the problem of site conditions and encouraged him to begin documenting them.

As the telling moves forward, Watts continues to valorize his role as citizen scientist by crediting his own powers of observation and reasoning with the discovery that problems existed at the USHCN stations. In his description of the first site location he visited, the Chico State University weather station, he expresses his surprise in discovering problems with the conditions of the station: "The first station had been converted to latex, but it also contained a surprise. . . . The NWS [National Weather Service] had installed the radio electronics just inches from the temperature sensor, *inside* the screen. Surely this station's temperature readings would be higher than the actual temperature of ambient air outside the screen" (5).

In this account of problem recognition, Watts's discovery is driven by his observations of conditions at the station as well as his native capacity for rational reasoning about the effects that these conditions might have on temperature measurement. Following this discovery, Watts further emphasizes his capacity, and the capacity of the citizen scientist, for empirical reasoning by refusing to jump to conclusions about the extent of the problem. Though he is concerned that the conditions at Chico University Station might be indicative of the whole USHCN, he suspends his judgment until he can collect further instances of problems by visiting other stations. Fortunately, his investigative plans included two additional station visits giving him the opportunity to gather more evidence. His observation of a second station in Orland, California, provided no evidence of the existence of a problem. However, conditions at the third station in Marysville, California, convinced him that poor siting conditions were likely endemic to measurement sites across the USHCN and may be introducing a warm bias into temperature measurements across the network. He writes of his experience at the Marysville station, "As I stood next to the temperature sensor, I could feel warm exhaust air from the nearby cell phone tower equipment sheds blowing past me! I realized that this official thermometer was recording the temperature of a hot zone near a large parking lot and other biasing influences including buildings, air conditioner vents, and masonry" (5). Stirred by the visceral physical experience of radiant heat around the sensor, Watts describes a moment of intellectual awareness and indignation: "Here we had an official climate-monitoring station, dubbed part of the 'high quality' USHCN network that provides data for use in scientific studies, actually measuring the temperature of a parking lot with air conditioners blowing exhaust air on it . . . !" (6).

In this moment of problem discovery, the citizen scientist literally stands alone. By erasing Pielke and other government scientists who had studied siting problems from his narrative, Watts highlights the perseverance of the citizen scientist's curiosity in discovering the problem and elevates the power of his rational/empirical reason over those of the institutional scientists who failed to identify or, worse yet, knew about and failed to disclose, problems with site conditions to the public. The latter failing not only valorizes the citizen scientist but also raises questions about the goodwill/credibility of the government's efforts to accurately report temperature data opening the door for charges of neglect. To maintain a strong neglect narrative in which poor siting conditions were knowingly tolerated by the government, it was necessary to deemphasize Pielke's previous advocacy to have the problem investigated as well his good faith efforts to engage with laypersons, such as Watts, to find solutions to the problem. Had these facts been presented, they would have challenged the notion that institutionalized science's monitoring or assessment of USHCN site conditions had been neglectful.

The absence of Pielke in Watts's introductory discovery narrative and the consequences of this absence—the valorization of the citizen scientist and promotion of a narrative of government neglect—tracks closely with frames in the conservative media and diverges from frames in the centrist/liberal media. This concurrence suggests that Watts's representation of the citizen-science project reflects and is likely informed by conservative perspectives. With the conservative media and, presumably, its audiences showing the most interest in the Surface Stations project, it's not surprising that Watts, a conservative himself, might have been inclined to remove Pielke from his problem discovery narrative to align his story with his audience and their perspectives. These inclinations were also likely reinforced by the fact that the publisher of the report was the Heartland Institute, a conservative think tank supporting policies skeptical of the severity of climate change or its source in anthropogenic activities (Heartland Institute, "Global Warming"). Watts's choice to present a problem discovery narrative that coincided with conservative frames, however, had repercussions for the capacity of citizen science to promote identification between laypersons and scientists. By embracing conservative frames, Watts obliterates any positive effects that might have resulted had he portrayed the relationship between himself and Pielke with more historical fidelity. Such a portrayal would have highlighted the supportive, collaborative relationship between the two and revealed their mutual interest in maintaining the integrity of the temperature measurement system. The fact that Watts chose not to make this kind of characterization illustrates how value commitments can shape the way that citizen science is represented in discourse and how these representations can impede potential improvements in identification between citizens and scientists.

Science before Scientists: Separating Scientific Method from Scientists and Scientific Institutions

Although the introduction of Watts's report obfuscates the role of Pielke and institutional climate science in the problem discovery narrative, thereby rejecting any identification with them, the main body of his argument embraces the methods, language, and style of climate science and scientists suggesting a move toward identification. This embrace seems to conflict with the strategy adopted in the introduction; however, Watts's recognition of climate science and climate scientists in the body of his argument remains perfectly aligned with the representations in the first part of the text if we consider two things. First, this shift in the discourse coincides with a change in the argumentative goals of the document. Whereas the purpose of the introduction is to favorably dispose the audience toward the arguer and his conclusions, the aim of the body of the argument is to make a compelling case for the validity of the findings of the Surface Stations project. Because the language, style, and methods of scientific argument carry authority, they are a means for Watts to establish the reliability of the conclusions of the Surface Stations project. The move from audience conditioning to argument, however, doesn't fully account for the sudden embrace of these features of scientific argument in the report. How can Watts use the methods developed by institutional science to validate his conclusions while at the same time maintain that these organizations have been negligent and incorrect in their assessments of these conditions? Overcoming this obstacle of argument requires a second consideration: that a distinction can be made between scientific language, style, and method and the agents and institutions of science that use them. In the body of his report, Watts relies on this distinction to avoid just such a conflict.

In the section "The Surface Stations Project," Watts signals the transition from the introduction to the main body of his argument by disclosing his collaboration with Pielke and his use of NOAA's criteria for judging station quality: "I worked with Dr. Pielke to encapsulate his survey methods into simple instructions any member of the public could understand and follow. . . . To rate the quality of the station siting characteristics, we used the same metric developed by NOAA's National Climatic Data Center [NCDC] to set up the Climate Reference Network" (8). In this transition, we can see Watts's efforts to separate the institutions and agents of science from its methods. Watts establishes this separation by characterizing his work with Pielke as strictly methodological leaving out any reference to the broader program of problem recognition Pielke had been involved in and its role in inspiring the project. This allows him to maintain his credibility as originator of the research but still associate the citizen-science project with climate science. Similarly, by focusing on the metrics of the NCDC rather than the fact that these metrics were created as part of the institution's effort to identify problems with site conditions, Watts is able to establish the credibility of the Surface Stations project's methods while maintaining a critical stance on institutional efforts to assess the effects of site conditions on temperature measurement.

In addition to drawing on the methods of climate scientists and scientific organizations to establish the scientific rigor of the Surface Stations project and its conclusions, Watts also adopts a scientific style of writing to support the credibility of the project. Unlike sections one and two, which are written in a largely colloquial, personal style, the content in sections three and four contain features of scientific discourse such as the persistent use of passive voice and technical terminology. In section three, for example, Watts explains how he used NOAA's classification system to rate the quality of the station sites. He writes, "The surveys are seen by dozens to hundreds of people, who readily point out errors or concerns, such as a misidentified station. In such cases where an error is identified, surveys are removed from the database,"

(Watts, *Is the U.S.*, 8). In these lines the use of the passive voice "are seen," "is identified," and "are removed" indicate a shift toward a scientific register that deemphasizes the role of the citizen scientist in the data-gathering process, a contrast to the valorization of the citizen scientist in the introductory section. This change in emphasis correlates with the change of argument stasis from arguing about the value of the project to the correctness of its methods and data.

In the fourth section, "Examples of Poor Siting," Watts's descriptions of his data-gathering activities resemble scientific writing even more closely, because he adopts not only the passive voice but also the technical terminology used in talking about temperature measurement. In one paragraph he writes, for example: "A trend illustrated by the photos above is for the newer style MMTS/Nimbus thermometers to be installed much closer to buildings and radiative surfaces than older Stevenson Screens. NOAA's sensor cable specification cites a maximum distance of ¼ mile, but installers often can't get past simple obstructions" (10). In these lines, "MMTS/Nimbus thermometers" and "Stevenson Screens" are proper nouns used by specialists to describe weather monitoring and data-collection devices. This kind of language would likely be unfamiliar to most nonspecialist readers as perhaps would be the technical phrase "radiative surfaces," which is a staple in scholarship on temperature measurement but not an otherwise commonly used phrase. These terms allow Watts to exhibit his expertise and affiliate himself with the specialty of weather monitoring and temperature measurement. At the same time, the adoption of technical terminology does not necessarily commit him to the social, political, and epistemological perspectives of the climate-science community.

Watts's use of a technical register continues into the "Findings" portion of the document in which he presents the reader with the data collected by the Surface Stations project. In describing his findings, he restates his method of investigation and offers precise quantitative evidence of the breakdown of station conditions: "Each Station has been assigned a CRN rating based on the quality rating system provided by NOAA. We found only 3 percent of stations surveyed meet the requirement of Class 1, while an additional 8 percent meet the requirements of Class 2. . . . Twenty percent of stations were rated as Class 3, 58 percent as Class 4, and 11 percent as Class 5" (Watts, *Is the U.S.* 16).

Up to this point, Watts's presentation of his methods, data, and results, though not perfectly commensurate with scientific standards of presentation, closely emulate the choices of language, style, and argument used by scientific reports. This carefully constructed technical style begins to unravel, however, as Watts discusses the conclusions drawn from the results of the citizen-science project. Whereas scientific writers cautiously limit their argument to the facts of the phenomena they investigate, Watts makes an intuitive leap in his conclusions from the quality of the sites, which had been rigorously established through the application of accepted scientific methods, to the quality of the data that the sites produce. Though the assumption that bad siting would create bad data seems reasonable from a nonspecialist perspective, from a scientific perspective no judgment about the quality of the data can be made without a careful statistical analysis of the relationship between the two. Such an analysis would necessarily include statistical evidence of a correlation between site quality and data quality as well as the type (warm or cool) and amount of bias introduced. Instead of relying on a direct statistical comparison of the data, Watts assumes that a relationship between site quality and data quality can be established using NOAA estimates of the degree of error a site's quality might introduce into its measurements. Using NOAA's calculations of the amount of error sites of different qualities could introduce, he concludes that because the amount of expected error in the system is much larger than the calculated rise in the temperature in the last century, the rise in US temperature must be artifact of these errors rather than a real environmental condition. He expresses this argument when he writes, "Recall that a Class 3 station has an *expected* error greater than 1°C, Class 4 stations have an *expected* error greater than 2°C, and Class 5 stations have an *expected* error greater than 5°C. These are enormous error ranges in light of the fact that climate change during the entire twentieth century is estimated to have been only 0.7°C. In other words, the reported increase in temperature during the twentieth century falls well within the margin of error for the instrument record" (Watts, *Is the U.S.* 16).

The disparity between Watts's understanding of the relationship between site quality and data quality and climate scientists' understanding of this relationship is illustrated not only by his assumption that measurement errors are equivalent to statistically identified biases but also by how he interprets the science he is using to defend his perspective. The latter problem is evidenced by his interpretation of the word *expected* in NOAA's rating system. In Watts's interpretation, "expected" means NOAA strongly believes that there is a certified relationship between a particular class of site quality and particular degree of error in the data. This interpretation is evident in Watts's italicization of "expected," which was meant to highlight the strength of the organization's commitment to these measures of error, which he believes supports his conclusion that global warming is an artifact of temperature measurement. From a scientific perspective, however, the term "expected" is full of vagaries and might be translated to mean, "We suspect but we just don't know because no disciplined assessment has been done." In fact, this interpretation of the term appears in the section of the scientific document where Watts draws his error values from. In the *Climate Reference Network (CRN) Site Information Handbook*, the scientific authors write, "The errors for the different classes are estimated values" (NOAA, *Climate* 5). The fact that these are estimated rather than precisely defined error ranges suggests that, from the perspective of climate scientists, they would be shaky warrants on which to base conclusions about global warming trends. Additionally, Watts fails to account for the fact that errors in measurement might be warmer or cooler. By automatically assuming that all errors reflect a warm bias, Watts opens his argument up to the kind of critique that a professional climatologist would have assiduously avoided.

Besides making unwarranted leaps in his technical argumentation between site quality and data quality, Watts also violates scientific conventions for argumentation by moving up the stases[18] from fact, to quality, to action. In "Accommodating Science," Jeanne Fahnestock explains that it is common for members of the lay public to extrapolate beyond scientific findings of fact and definition to raise questions or make arguments at the stases of cause, quality, and action. However, scientists resist these kinds of leaps in their work (Fahnestock 245–46). In the final section of the report "Policy Implications and Recommendations," Watts crosses the typical statial boundaries of technical argument when he makes qualitative attacks on the practices of governmental organizations responsible for studying climate change and the government's climate-change policy. In the first paragraph of the section, he opens with the bold conclusion at the stasis of quality: "The USHCN stations are not sufficiently accurate to use in scientific studies or as a basis for public policy decisions" (Watts, *Is the U.S.* 17). He also concludes that by producing or associating themselves with this bad data, and the bad practices used to generate it, none of the government organizations associated with climate change should be trusted. After listing all of the major national and international agencies involved in studying climate (NOAA, NASA, NCDC, IPCC) as well as their managers, Watts writes, "The findings and recommendations of these highly respected and influential scientific and political organizations are now in doubt. The data currently used . . . are unreliable. The truth of that claim can be established only with new and more-reliable data" (17). This qualitative conclusion reconfirms Watts's strategy of separating scientific methods for establishing the credibility of temperature data from the validity of organizations that developed them. The only way to know that these institutions have erred after all is to have a correct scientific method for distinguishing what is in error and what is not. NOAA, NASA, and the NCDC provide the methodological means by which their own errors are brought to light; however, they are dismissed as having been incautious in their conclusions about climate change and, therefore, unreliable as credible sources on the subject.

Following his arguments at the stasis of quality, Watts moves to the final stasis of action by recommending the kinds of scientific methods NOAA and the NCDC should undertake to improve their work. In this list he includes the following suggestions: "A pristine data set should be produced . . . to quantify the total magnitude of bias." And "NOAA should undertake a comprehensive effort to improve the siting of the stations and correct the temperature record for contamination" (Watts, *Is the U.S.* 17). What is interesting about these recommendations is that they propose corrective measures that NOAA and the National Climatic Data Center (NCDC) had already suggested and adopted. Watts's mention of them further supports the interpretation that he is attempting to separate the agents and organizations of climate science and their methods. Though NOAA and the NCDC have lost all credibility, in Watts's estimation, their methods for improving temperature measurement are sufficiently credible for him to recommend as solutions to the problem of inaccuracies in temperature measurement. By using these already agreed-upon methods but not citing their sources, Watts is able to present scientifically sanctioned solutions to the problem of temperature measurement while at the same time giving the audience the impression that these solutions arise from his citizen-science experience rather than from the institutions of science he seeks to discredit.

This close reading of the introduction, body, and conclusion of Anthony Watts's report *Is the U.S. Surface Temperature Record Reliable?* reveals a rhetorical balancing act between drawing on climate science and scientists to credential the work of the Surface Stations project while at the same time casting aspersions on these same institutions for failing to properly recognize the shortcomings of their data collection process. To establish the credibility of his critique, Watts relies on Pielke's work and the language, style, and methods of climate science. At the same time, he avoids crediting climate scientists or climate-science institutions for having developed these solutions by erasing them from his problem discovery narrative and separating the institutions of climate science from their methods of problem assessment in the final sections of his report.

By examining the discourse and argument created by Watts about the Surface Stations project, it is clear that though citizen science may foster collaborations of mutual benefit and interest, these collaborations may not always translate into public discourse and argument that highlights the potential of citizen science to improve relationships and increase identification between laypersons, scientists, and scientific institutions. Although Watts is the sole author of the report and, therefore, responsible for its contents, it is also worth considering whether an altogether more informed and less strongly politicized discourse might have been created if he had received more editorial input from Roger Pielke Sr. during the report's creation. In an email interview with Pielke, the climate scientist stated that he had never looked at a draft of the report or gave Watts any advice about it (Pielke, interview).[19] Had

Pielke involved himself in the production of the discourse and argument in the report, he might have encouraged Watts to consider alternative choices of style and content. He might, for example, have insisted on changing the problem discovery narrative to recognize his own work and the work of the institutions of climate science as influences on Watts's efforts, thereby highlighting the mutually supportive role scientists and citizens could play in recognizing and addressing the shortcomings of climate science. He might also have identified Watts's overextension of his arguments about the relationship between the quality of sites and the quality of data. If citizen-scientist collaborations are to foster the development of better understandings and relations between laypersons, science, and scientists, then it is essential, as Anthony Watts's report illustrates, for lay participants and scientist alike to be cooperatively involved in the creation of public discourse and argument about their citizen-science endeavors. As part of their participation, they should consider how their values as well as the values of their audience might influence how citizen science is represented. By more carefully considering this rhetorical dimension of citizen science, they might better negotiate how, collectively, they represent their work in the public sphere.

Surface Stations in the Technical Sphere

In addition to considering citizen science's capacity (or lack thereof) to promote mutual understanding between laypersons and scientists in the public sphere, this chapter is also interested in exploring the previously unexamined question, Can citizen science provide access for laypersons to technical sphere argumentation? The assessment of the scientific literature that follows reveals that in fact Anthony Watts's report spawned quite a bit of discourse and argument in the technical sphere, some of which recognized him for his efforts to assess site conditions and even credited him as a coauthor of scientific discourse. Though this evidence provides a simple yes to the question of access, it invites a more complex and interesting query: Can the socio-political and institutional values and goals of science affect the way that citizen science is characterized in technical sphere discourse and argument? This section explores this question by examining 1) the context in which scientific discourse and argument about the Surface Stations project emerged, 2) the critiques of Watts's report by representatives of leading scientific institutions, and 3) the responses to these critiques made by supporters of the Surface Station project's empirical conclusions.

To understand the character of the reception of the Surface Stations project in the technical sphere, it is essential first to examine the context in which discourse and argument critical of the project emerged. This involves identifying groups in the technical sphere that might be critical of the project, and understanding what might compel them to respond to it. Previous analysis in this chapter identifies three technical sphere institutions that were the most heavily involved in assessing the problem of temperature measurement: NOAA, NASA, and the NCDC. Of these three, the National Climatic Data Center (NCDC) had the greatest investment in the issue. The NCDC is the federal organization tasked with gathering, storing, and analyzing data about climatic conditions in the United States. Toward this end, it maintains a number of networks of weather stations across all fifty states including the US Historical Climatology Network (USHCN). Because the NCDC is the premier organization for gathering land surface temperature data and because it is the longtime custodian of the USHCN, it is not surprising that representatives of this organization paid close attention to the Surface Stations project and had exigence to respond to it.

From the very launch of the project, the NCDC's communications and activities indicate that Surface Stations was on their radar and that they were prepared to respond to what they believed was a threat to the credibility of their organization from the project. On June 19, 2007, only two weeks after Surfacestations.org went live, the director of the NCDC Thomas Karl sent an email to Phil Jones, a colleague at the center, in which he wrote, "We are now responding to a TV weather forecaster [Anthony Watts] who has got press. He has a website of 40 of the USHCN stations showing less than ideal exposure. He claims he can show urban bias and exposure biases. We are writing a response to our public affairs. Not sure how this will play out" (qtd. in Revkin).

On June 25, a few days after this email, the NCDC responded to the citizen-science endeavor by shutting down access to what had previously been a public list of locations and addresses of the stations in the USHCN (Watts, "NOAA/NCDC"). Because many of the stations were located on private residences, the NCDC argued that they had a duty to protect the privacy of their observers. Without access to the station locations, Watts and the citizen-science observers he had recruited could no longer document station conditions. After blogging about the NCDC's censorship, Watts received information from one of his online followers that NOAA and the NCDC had made the addresses and names of stations available in other venues. A week later on July 7 he received notice from NOAA that after legal consultation they had restored access to the names of the observers in charge of the stations (Watts, "NOAA and NCDC Restore"). With this information Watts and his contributors could continue their documentation work.

Although the NCDC's initial response to the Surface Stations project was to block access to observer information, its subsequent strategy for dealing with the potential public relations threat it posed was less confrontational. It decided to embrace the project and make an effort to educate Anthony Watts

on the science of temperature measurement and the NCDC's efforts to ensure reliable climate data. In February of 2008, the director of the NCDC
Thomas Karl extended an invitation to Watts to present the methods and findings of his project at their headquarters in Asheville, North Carolina. In late
April 2008, Watts paid the center a visit to have an "exchange of ideas and
information." During his time in Asheville, Watts met with the center's top
managers and scientists, gave a talk about the Surface Stations project, and
toured a few of the center's new Climate Reference Network measurement
sites (Watts, "Road Trip" and "Day 2"). During the visit, members of the organization listened to Watts's concerns and even expressed their appreciation
for what Watts was trying to accomplish with the citizen-science project. The
trip seemed to be a remarkable change of attitude by the organization that
had, less than a year before, attempted to shut down the project. This transformation is reflected on by Watts in his blogs about the trip. He remarks,

> I want to extend my heartfelt thanks to . . . the entire CRN science team . . .
> for answering all my questions and taking such careful time with me. Addi-
> tionally, I wish to thank Dr. [Thomas] Karl [Director of the NCDC] and As-
> sistant Director Sharon LeDuc for hearing my concerns and offering ideas.
> Everyone there at NCDC made me feel welcome and appreciated. ("Day 2")

Watts's visit to the National Climatic Data Center headquarters marked a
rare if not unprecedented show of acceptance by a large and important scientific governmental organization of a citizen-led citizen-science project. It
might also be considered the high point in the relationship between Watts
and the NCDC. With the publication of Watts's report, relations between the
citizen scientist and the NCDC lapsed into confrontation once again. Less
than three months after the release of *Is the U.S. Surface Temperature Record
Reliable?* representatives of the NCDC Matthew Menne, Claude Williams,
and Michael Palecki submitted for review the paper "On the Reliability of
the U.S. Surface Temperature Record."[20] In the paper, the authors challenge
Watts's scientific conclusions about the relationship between the quality of
site conditions and the quality of temperature measurements. At the center
of this challenge are the scientific questions, What degree and type of bias is
introduced by site conditions into temperature measurements? and To what
extent is qualitative data about site characteristics helpful in understanding
these biases? An examination of the technical sphere arguments launched
by representatives of the NCDC suggest that though they advanced legitimate technical critiques of Watts's work, they also extended their criticism
of the Surface Stations project into the realm of value argument. The question the analysis in the next section endeavors to answer is, Why? Why would

the NCDC stretch beyond a scientific critique of Watts's work to challenge it at the stasis of value?

NCDC and the Value of Statistics

Perhaps the most suitable explanation for why Menne, Williams, and Palecki were compelled to stretch their argument is because they could not easily dismiss the value or legitimacy of the Surface Stations project's site assessment by dismissing its conclusions about measurement biases. In their paper the authors reveal their awareness of the social and technical legitimacy of the project by the way they strike a balance between praising the project's efforts to document site conditions and challenging Watts's conclusion about the relationship between site quality and data quality. Evidence of the authors' support for the project's methods for documenting site conditions and ranking site quality appears in the methods section of the paper. Here the authors affirm their acceptance not only of the project's method for rating sites but also its general conclusions about the quality of sites by using them as the basis of their own investigation: "The exposure characteristics of a subset of USHCN stations have been classified and posted to the web by the organization surfacestations.org. . . . To evaluate the potential impact of exposure on station siting, we formed two subsets from the five possible USCRN exposure types assigned to the USHCN stations by surfacestations.org" (Menne, Williams, and Palecki par. 6).

In addition to accepting the project's methods and results, the authors go out of their way to recognize the social importance of the project by praising Watts and the work of his volunteers for gathering data on station quality. In an "acknowledgment" footnote at the end of the paper, they comment, "The authors wish to thank Anthony Watts and the many volunteers at surfacestations.org for their considerable efforts in documenting the current site characteristics of USHCN stations" (Menne, Williams, and Palecki fn 21).

Although there is praise for the citizen scientists for participating in data gathering and acceptance of their methods, the authors take issue with Watts's argumentative leap in his report from the quality of the sites to the quality of the temperature data of those sites. This argument is introduced in the first paragraph of the paper where the authors write, "*Watts* [2009], in particular, has speculated that U.S. surface temperature records from the USHCN from the last 30 years or so are likely biased high (warm) thereby artificially enhancing the magnitude of the observed temperature trends" (Menne, Williams, and Palecki par. 2). In this characterization of Watts's conclusion, it is notable that the authors use the term "speculated" to suggest that Watts's argument about the link between station quality and data quality is guesswork unsupported by a rigorous investigation. The goal for Menne, Williams, and

Palecki in their paper is to establish whether such a connection actually exists and, if so, then how much and what type of bias it might introduce into the temperature record.

In order to assess the existence and extent of possible bias, the authors divide up the stations by their designated classifications and statistically compare them. Using the Surface Stations project's rankings they split the stations into a "good" group that includes well-sited Class 1 and 2 stations and a "bad" group that includes poorly sited Class 3–5 stations. Next, they use a geographically representative sample of stations to compare "good" and "bad" stations across the United States in similar regions to find out just how much and what kind of bias (warmer or cooler) existed between them. They do this both with the unadjusted raw data from USHCN temperature measurements and with data that had been statistically homogenized, or adjusted for known errors.[21] They conclude that in the statistically *adjusted* data there was almost no difference between the data from "good" and "bad" stations. In other words, statistical adjustments had corrected for any biases that might be the consequence of poor station siting so that they had no effect on the historical temperature record. In their comparison of the *unadjusted* raw data from "good" and "bad" stations, however, the authors found, surprisingly, that there was a cool bias in maximum temperatures. In other words, stations sited near parking lots, air conditioners, and BBQs had, on average, cooler readings than their better-sited counterparts: a counterintuitive rebuttal to Watts's apparently commonsensical conclusions that these sorts of phenomena would introduce a clear warming bias.

Although Menne, Williams, and Palecki's statistical arguments present a technically compelling correction to Watts's conclusions about the link between station quality and data quality, the authors do not let their argument rest solely on these mathematically derived conclusions. Instead, they move beyond arguments from quantity to arguments about the value of their quantitative method. This change of strategy in the conclusion of the paper suggests an effort by the authors to rhetorically address the lingering socio-political and epistemological challenges presented by Surface Stations project—the fact that the majority of stations were found to be poorly sited and that by uncovering this fact the Surface Stations project had raised questions about the credibility of the NCDC. Because the authors cannot challenge the essential correctness or appropriateness of the Surface Stations project's methods, they have to find a way to diminish the value of these methods and elevate the value of their own. Toward this end, they contrast their quantitative approach to the project's qualitative one. In the paper's conclusion, the authors explain, "Given the now extensive documentation by surfacestations.org [*Watts,* 2009] that the exposure characteristics of many USHCN stations are far from ideal, it is reasonable to question the role that poor exposure may have played in biasing . . . temperature trends. However, our analysis . . . illustrate[s] the need for data analysis in establishing the role of station exposure characteristics on temperature trends no matter how compelling the circumstantial evidence of bias may be. In other words, photos and site surveys do not preclude the need for data analysis" (Menne, Williams, and Palecki par. 18).

In these lines, the authors frame their statistical conclusions as a cautionary tale of what can happen when you trust too much in experience and too little in mathematical reason. They point out that while the circumstantial photographic evidence of site conditions certainly suggests that poor siting would introduce a false warming bias into temperature data, this assumption was mistaken. Through statistical data analysis, however, it is possible to check and correct these misperceptions. It is, therefore, these quantitative conclusions that we must accept no matter how contradictory they seem to our experience. As to how the experience and common sense intuition of citizen-science observers could have gotten it so wrong, the authors suggest only that "the reason why station exposure does not play an obvious role in temperature trends probably warrants further investigation" (8). What is certain, though, is that while the Surface Stations project had some value for science, this value was ultimately limited by its method. By placing the *loci of quantity* above the *loci of quality*,[22] representatives of the NCDC maintain the value of their specialized scientific work trumps the experience and common sense intuitions of Watts and his citizen-science researchers.

In Defense of Qualitative Assessment

Though representatives of the NCDC defended their credibility and expertise by challenging the value of the Surface Stations project's qualitative approach to assessing measurement bias, their arguments were countered by other participants in the technical sphere willing to defend the value of the project's qualitative approach. A year after the publication of Menne, Williams, and Palecki's paper, Fall and coauthors' article "Analysis of the Impacts of Station Exposure on the U.S. Historical Climatology Network Temperatures and Temperature Trends" (2011) was published in the *Journal of Geophysical Research*. In this article, which included both Pielke and Watts as coauthors, the authors present their own statistical examination of the data from the Surface Stations project using a large quality-controlled sample from 82.5% of the sites assessed (Fall et al. par.1). In the paper they both confirm and complicate the findings of Menne, Williams, and Palecki while defending the importance of qualitative site investigations and, thereby, the value of the citizen science.

In the "Introduction" section of the document, the authors open their argument by mounting a defense of the Surface Stations project's qualitative investigation by highlighting its value in ascertaining the sources of bias in measurement and correcting the potential errors of quantitative assumptions about bias:

> Overall, considerable work has been done to account for [statistical] inhomologies and obtain adjusted data sets for climate analysis.
> However, there is presently considerable debate about the effects of adjustments on temperature trends. . . .
>
> [The Surface Stations project's] photographic documentation has revealed wide variations in the quality of USHCNv2 station siting. . . . It is not known whether adjustment techniques satisfactorily compensate for biases caused by poor siting. (Fall et al. par.4, par.9)

In the first of the two paragraphs quoted here, the authors set up the exigence of their research project by arguing from the *topos* of uncertainty[23] that there is "considerable debate" about the statistical adjustment in temperature trends. The uncertainty in the statistical treatment of the data, in their estimation, is a consequence of the general lack of qualitative knowledge about how siting factors influence temperature trends. Thanks to the work of Watts and his citizen-science volunteers, however, it is finally possible to know the extent of an important source of statistical biases, site conditions, and quantitatively test their influence on temperature data. By setting up qualitative knowledge of siting factors as a precondition for robust quantitative descriptions of measurement biases, Fall and coauthors change the order of qualitative and quantitative analysis in the value hierarchy of scientific methodology. They argue essentially that because the truth or value of statistical claims is predicated upon qualitative knowledge about station conditions this qualitative knowledge has a higher value. The "garbage in equals garbage out" argument here draws from the commonplace of cause and consequence in which, as Aristotle explains, "One thing may be shown to be more important because it is a beginning and another thing is not" (I vii 17).

After asserting the value of a qualitative approach over a quantitative one, the authors begin their investigation of whether and in what manner siting factors bias temperature measurement. In pursuing this question, they use both similar and slightly different approaches than Menne, Williams, and Palecki do in their assessment. Though it is rare in science that the same analysis or experiment is replicated, the stakes of these investigations are high—the validity of temperature measurements and, therefore, climate change policy hang in the balance. Like Menne, Williams, and Palecki, the authors divide and compare the adjusted and unadjusted data from "good" and "poor" stations. Their results support the counterintuitive findings of Menne and his coauthors of the existence of a "cool" bias in poorly sited station data: "[Our analysis] largely confirms the more limited findings of *Menne et al.* [2010] that poorer-sited stations produce larger minimum temperature trends and smaller maximum temperature trends" (Fall et al. par. 36). Unlike these authors, however, Fall and coauthors include substantially more and better quality-controlled data in their assessment[24] and pursue a broader set of investigations on the influence of site conditions on measurement biases. In addition to comparing the measurements from "good" and "bad" stations, for instance, the authors also compared the diurnal temperature trends (the difference between maximum and minimum daily temperatures) to assess whether station sitings bias this measure of temperature. They discovered that although most station types (CRN 1–4) exhibit no significant difference in the measurement of diurnal temperature trends, poorly sited stations (CRN 5) had a clear statistical bias toward an artificial narrowing of the difference between daily minimum and maximum temperatures. Although this finding has no direct significance for the climate change debate, it has important implications for Pielke and Watts's defense of their citizen-science project. First, it suggests, unlike Menne, Williams, and Palecki's paper, that sensor siting has a direct impact on temperature measurement, which has not been adequately accounted for in the adjustments of the statistical data. They explain, "The homogeneity corrections are not as successful in adjusting diurnal temperature range" (Fall et al. par. 43). By identifying this hole in the current program to statistically homogenize data, the authors show that site conditions matter and that, as a consequence, qualitative efforts to document and rank conditions at sites matter. They explain, "The evidence presented in the preceding section supports the hypothesis that station characteristics associated with station siting quality affect temperature trend estimates" (par. 48).

By examining and comparing scholarly work critiquing and defending the Surface Stations project in scientific literature, a number of observations about how citizen science is received and handled in the technical sphere can be made. The first observation is that access of citizen science to the technical sphere may be influenced by the extent to which citizen-science projects adopt the methods and practices of the specialized science(s) related to the phenomena that they are assessing. In the case of the Surface Stations project, because Watts worked closely with Pielke to design or identify rigorous methods for documenting and rating stations, these methods were deemed acceptable within the community of specialists and adopted by all participants in the technical sphere conversation. Secondly, I observe that the efforts of citizen scientists were considered valuable to participants in the technical sphere. This was evidenced by the praiseful acknowledgement of the project even in articles that were critical of it. However, despite general acceptance of the citizen science's value and the appropriateness of its methods and data, we can observe, thirdly, that this was insufficient in the technical sphere to gain the acceptance of the conclusions drawn by citizen scientists. In the case of the Surface Stations project, neither critical nor supportive scientists accepted the conclusions in Watts's report that data gathered by the project revealed a warm bias sufficient to account for warming in decadal and centennial temperature trends. Rejection of Watts's thesis suggests that though a nonspecialist citizen scientist might gain access to and recognition in a specialist debate, they are still subject to the conventions for evidence and argument of the field in which the debate takes place. In this case, Watts's use of error ranges estimated for the various site rankings to account for the centennial trend in climate change, and his assumption that the biases introduced by poor siting were always warm biases, conflicted with statistical practices used within the specialist community.

The fourth and, perhaps, most significant observation is that like lay representations of citizen science, expert representations are influenced by sociopolitical and institutional values and goals. By examining the context in which technical sphere arguers engaged with the citizen-science project, it was revealed that the NCDC considered Surfaces Stations a threat to its public credentials as a reliable source for data about temperature and climate change. As a response to this threat, the NCDC made value arguments that elevated their quantitative methods for assessing the biases of temperature measurement over the qualitative methods pursued by the citizen scientists. In response Pielke and other supporters of the citizen-science project defended the value of the qualitative observation. The role of the *loci of quantity* and *quality* as markers for scientific and lay argumentation has been broadly discussed by scholars in rhetoric and environmental justice studying *public sphere* debates between specialists and laypersons.[25] However, this analysis reveals significantly that this division can also be part of *technical sphere* argument. This similarity suggests that socio-political and institutional values penetrate into the technical sphere and affect the way that citizen science is characterized in discourse and argument.

CONCLUSION

The goal of this chapter has been to investigate whether and to what extent citizen science might influence the relationship between laypersons, science, and scientists in the public and technical spheres. Current research by scientists, scientific educators, and sociologists of science has suggested that cocreative citizen-science projects promote learning and identification between laypersons and scientists. This chapter has both tested these conclusions arrived at by PPSR and PES scholars and extended their investigations. It has shown that though the Surface Stations project brought a climate scientist and a climate change critic into a cooperative, collaborative engagement and gave the latter access to the technical sphere debate over temperature measurement, the discourse and argument generated from this engagement did not in all cases promote or reflect the collaborative spirit of the enterprise or encourage greater identification between laypersons, scientists, and scientific institutions. Close textual analysis of Watts's public sphere argument revealed, for example, that his personal convictions as well as the character of his audience encouraged him to create a narrative of problem discovery that valorized the citizen scientist and cast aspersions on climate science by leaving out important contributions of government scientists and scientific institutions to the study of siting problems. A similar analysis of the context and content of argument generated by climate scientists who responded to the Surface Stations project showed how scientists at the National Climatic Data Center used value arguments to attack the citizen-science project and protect their reputation as a reliable source for data about temperature and climate change.

These findings suggest that citizen science is subject to and shaped by the environment of argument in which it exists including the personal, social, and political values and goals of its participants. Because of its rhetorical character, citizen science requires, therefore, more than just mutual interest in a particular natural phenomenon and active collaboration by laypersons and scientists in studying it to promote improved identification and understanding between these two groups. Achieving these outcomes requires an active assessment of and engagement with the values and contexts of knowledge production for both citizens and scientists. Though this assessment/engagement may take a variety of forms, this chapter has endeavored to show how it might look from a rhetorical perspective. By working collaboratively with one another, or in conjunction with a rhetorical "honest broker,"[26] both parties might develop better self-awareness of their commitments, how those commitments manifest themselves in their discourse and argument, and how they shape the public or technical sphere debate in which they play a role. By attending to these rhetorical dimensions of discourse and argument, citizen science might yet realize its potential as a critical tool for developing better relationships between citizens, science, and scientists.

5

A Tale of Two *Logoi*

Citizen Science and the Politics of Redevelopment

The previous chapters in this book have discussed the role of digital-age citizen science in the development of citizen-centered communication, its place in the transformation of the ethos of grassroots groups, and its part in the development of relationships between scientists and laypersons. Though all of these chapters have touched in some way on issues of public policy and argument, none of them has specifically pursued the question, To what extent and in what ways might Internet-enabled citizen science influence public policy arguments and outcomes? This final chapter examines this question by exploring a policy debate over urban development and planning in the East London borough of Lewisham in which citizen-science activities driven by technical exigencies played a central role. To appreciate the influence of citizen science on policy arguments made by members of the public,[1] the first section of this chapter explores the *logos* developed by citizen scientists to make the case for the removal of a neighborhood scrapyard. To understand whether, and if so how, these arguments influenced policy outcomes, the second section examines how local policymakers responded to and used the results of the citizen science to promote and achieve policy goals. By exploring this case in detail, the chapter illustrates how digital-age citizen science can be an inventional resource for policy investigations and can play an important role in creating common ground for policy argument by integrating the nonexpert *logos* of lived experience with elements of the expert *logos* of techno-scientific rationality. Finally, it illustrates how contextual factors, in particular political agendas, can shape the way in which citizen science influences policy argumentation and action.

DIGITAL-AGE CITIZEN SCIENCE AS INVENTIONAL RESOURCE

As the cases in the previous chapters have shown, the Internet and Internet-connectable devices can play a significant role in citizen science by facilitating the development of novel forms of communication, opening new lines of argument, and creating new spaces in which citizens and scientists can interact. In this chapter I examine how digital technologies are also sources of invention, particularly in the development of citizen-science projects aimed at public policy interventions. Digital technologies serve as sources of invention for policy-centered citizen science in the sense that citizen-science projects can grow out of an academic interest in finding ways to use existing digital technologies to address policy issues. The inventional power of digital technologies has been recognized by scholars in geography and urban planning. For example, in the opening lines of a 2006 literature review on the emerging field of Public Participation Geographic Information Systems (PPGIS), geographer Renee Sieber writes, "It is an odd concept to attribute to a piece of software the potential to enhance or limit public participation in policymaking, empower or marginalize community members to improve their lives, counter or enable agendas of the powerful, and advance or diminish democratic principles. However that is exactly what is happening with geographic information systems (GIS), the social application of which has captured the attention of researchers in diverse disciplines, including urban planning, law, geography . . . and conservation biology" (491).

As these lines suggest, one of the central inventional questions driving researchers interested in applying new digital technologies to policy is, Can digital technologies be used to empower community members to improve their lives and counter agendas of the powerful? Some researchers have sought to answer this question through the development of digitally supported citizen-science projects. In the United Kingdom, for instance, researchers at the University College London (UCL) have formed the ExCiteS research group—Extreme Citizen Science with some of the letters removed—to "develop methodologies that enable communities . . . to ask research questions and collect and analyze data to advance local interests" (Rowland). The citizen science described in this chapter is one of the research group's many projects aimed at pursuing this goal. It took place in the London borough of Lewisham, which is located in the east and south region of the city—a region that includes some of the city's most economically depressed populations. Pepys Estate/Deptford, the neighborhood of Lewisham borough that was the specific site for the project, is characterized by one policy analyst in the following fashion: "While certainly very much tidier than 20 years ago, Evelyn and Newcross wards [the former of which includes Pepys Estate/Deptford] remain within the most deprived 10% of English neighborhoods" (Potts 11).

Local and national governments in the United Kingdom have taken measures to help rejuvenate these economically depressed parts of London through an ambitious program of urban planning and development. One prominent example of this redevelopment program was the siting of the majority of the venues for the 2012 Olympic Games in six eastside boroughs[2] (Great

Britain, Dept. of Culture). Less well-known but equally important has been the Thames Gateway project,[3] a redevelopment program aimed at introducing more modern, environmentally friendly buildings and recreational spaces to many of the inner-east and south London boroughs abutting the Thames. This program has included collaborations between national and regional governments, nonprofits, and universities to encourage a holistic approach to redevelopment. One of these collaborations, Mapping Change for Sustainable Communities,[4] was part of UCL's ExCiteS research program. It was funded by the UrbanBuzz initiative and included partnerships between faculty and students in UCL's civil, environmental, and geomatic engineering departments; members of the nonprofits London 21 and London Sustainability Exchange; and participants in the regional planning organizations London Thames Gateway Forum and Planning Aid for London ("Making Maps Work").

The goal of the Mapping Change for Sustainable Communities project was to use the Internet and digital mapping tools to help local communities impacted by the Thames Gateway development to "understand how any proposed changes may affect them and to feel confident about making their voices heard" in response to these changes ("Mapping for Sustainable Communities," *UrbanBuzz* 1). To empower these communities, the academics and NGOs associated with the project collaborated with community groups to develop digital maps and online forums. Through these maps and forums, community members could represent concerns about their localities (crime, blight, pollution, etc.), or celebrate features of their community they wanted people to know about (history, festivals, community meetings, etc.). Working with the nonprofit London 21, researchers at University College London identified five partner sites[5] for their project including Pepys Estate in the borough of Lewisham ("Pilot Groups"). Once the sites for the mapping projects were identified, workshops—developed by Muki Haklay, professor of Geographic Information Science (GIS) at UCL, and Louis Francis and Coleen Whitaker of London 21—were held in each community to identify what kinds of digital mapping projects residents were interested in doing. In the case of Pepys Estate/Deptford, residents had made complaints to their borough council for years about the noisy operations of a local scrapyard. At the behest of the borough councilor, Haklay and his team were asked to work with residents to help them collect data and create digital maps of the noise levels in their neighborhood (Pepys Community 8).

Like the radiation measurements of Safecast, the community sound-mapping project required the use of technical instruments, sound meters, which participants had to learn how to read and operate. These activities exposed them to technical information about noise thresholds and methods of acoustical measurement that were typically the domain of the technical sphere. Though neither Haklay nor the researchers at London 21 were specialists in acoustics, their technical/mathematical background in Geographical Information Science (GIS) and their skills as academic researchers helped them develop a reasonably robust understanding of the basic methods of acoustical assessment and how to pass it along to participants (Haklay, Interview). Based on their research, Haklay and his team designed a protocol for gathering information about noise levels and materials for teaching that protocol to the community members engaged in the sound-mapping project (Haklay, Francis, and Whitaker 27).

The basic protocol they developed required participants to take noise readings three times a day over a period of seven weeks.[6] In each instance of recording, the measurer noted the time and date of the session on a special form.[7] They also marked their location on a paper map. Once all of this basic information was recorded, they took three sound readings in three one-minute intervals to produce an average maximum noise level for their location ("Noise Mapping Toolkit" 7). In addition, they circled words on their recording sheets that described the quality or the intensity of the sound (e.g., silent, deep, constant, random, enjoyable, disturbing, etc.) and noted what the loudest source of sound was at the time of the recording. Once completed these forms were collected at the end of the observation period and checked for accuracy and consistency by the UCL/London 21 team. The vetted readings were then entered into a database and loaded onto geographic information systems software to make maps of the survey results. Some of the maps created through this process included a grid of squares digitally overlaid on a map of the neighborhood. Each of the squares contained a number indicating how many measurements were taken in each square. The noise levels in each square were communicated using shades of red, with lighter shades of red indicating low noise levels and darker shades of red indicating higher levels (figure 9). In addition, Haklay and his team created digital maps without grids that used dots to show precisely where measurements were taken around the area (figure 10). They then color-coded these dots to correspond to the locations at which a particular source of sound (the scrapyard, traffic, airplanes, etc.) was considered by the measurer to be the loudest sound.

Though the production of digital maps was one of the primary tasks of ExCiteS' Mapping Change for Sustainable Communities project, another major responsibility was to facilitate dialogue about the maps between the citizens who had created them and the local government officials who might be able to take action on the problem being mapped. What makes the Pepys Estate/Deptford noise mapping project rhetorically interesting is that in this dialogue the data, maps, and experiences of the citizen scientists became important catalysts for developing policy arguments about the levels of noise in the environment and what actions should be taken to respond to them. An examination of the policy arguments developed from this project illumi-

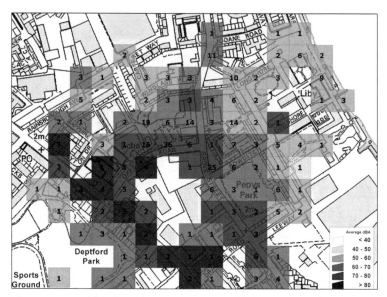

Figure 9. Sound levels on Pepys Estate. (Used by permission from Muki Haklay; see Whitaker.)

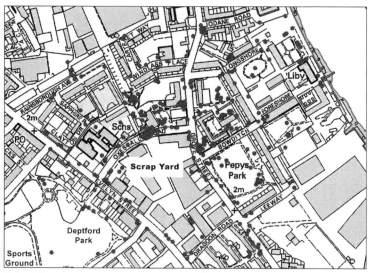

Figure 10. Sound reading locations on Pepys Estate. (Used by permission from Muki Haklay; see Whitaker.)

nates the power of citizen science—driven by the technological exigencies of finding policy applications for digital mapping systems—to bring the *logoi* of policymakers and community members closer together. However, it also shows that the relationship between community argument and policy action based on citizen science are not necessarily straightforward.

LITERATURE REVIEW: *LOGOS* AND POLICY ARGUMENT

Before examining the relationship between citizen science and policy argument in the Pepys Estate/Deptford case, it is important to identify the rhetorical focus of this investigation and the scholarship in rhetoric, sociology, and public policy that are relevant to thinking about citizen science in the context of policy argumentation. Whereas the rhetorical foci in previous chapters have included visual argument, *ethos*, audience, and context, this chapter focuses on *logos*. In particular, it examines how citizen science influences the *logos* of community action groups in their policy arguments. According to Aristotle *logos* is "the proof, or apparent proof, provided by the words of the speech itself" (I ii 1356a). This proof includes the facts or evidence of a case and the strategies of argument used to relate the facts to one another. In public policy arguments—arguments Aristotle labels political or deliberative rhetoric—the general goal of appeals to *logos* is to make the case for future action by establishing "the expediency or the harmfulness of a proposed course of action" (I iii 1358b). This kind of argument stands in contradistinction to *forensic* (legal) argument, which focuses on past fact to make the case for the justice or injustice of action, and *epideictic* argument, which bestows praise or blame on a subject according to present socio-cultural values.

Though Aristotle makes some very useful distinctions between political rhetoric and other rhetorical genres, recent scholars have been more circumspect about categorizing argument strictly by the division of past fact, present values, and future action. In *A Rhetoric of Argument*, for example, Fahnestock and Secor point out in their discussion of the stases, or resting places of argument, that arguments at the stasis of proposal, the typical stasis for policy argument, tend to include arguments based on fact, cause, and value.[8] They explain, "Proposal arguments tend to follow a predictable pattern: they first convince an audience that a problem exists, and then they propose a solution. To achieve their goal, they build on all the previous types [of stases] we have examined [definition, cause, and evaluation]" (Fahnestock and Secor 285). For the purposes of this chapter, I will maintain Aristotle's definition of deliberative argument as argument whose goal is to establish the expediency or harmfulness of a proposed course of action, but adopt Fahnestock and Secor's more comprehensive perspective on the roles of facts, causes, and values in the *logos* of policy argument.

Because citizen science, as this book defines it, is a relatively recent phenomenon, there is, to my knowledge, only a single publication in the fields
of rhetoric, public policy, or sociology that has examined the influence of
citizen science on the *logos* of policy argument. This publication, Gwen Ottinger's "Buckets of Resistance," will be discussed later in this section. Before examining Ottinger's work, however, it is useful to consider more generally what the trend in rhetoric and communication scholarship has been
in thinking about the *logos* of policy arguments at the intersection of science,
technology, and public policy. Much of the discussion in this literature has
been devoted to defining or making distinctions between the techno-scientific
logos of institutional authorities and the *logos* of nonexpert citizens (Kinsella
84). For scholars engaged in this conversation, the *logos* of nonexperts is typically grounded in experience either from social interaction or the lived experience of material conditions (Irwin 3; Fischer 44; Leach and Scoones 18–
21). This characterization is highlighted, for example, in policy theorist Frank
Fischer's work *Citizens, Experts, and the Environment* in which he makes the
following comment about the logical resources of local constituencies: "[Lo-
cals] typically possess empirical information about the situation unavailable
to those outside the context. While such local knowledge cannot in and of
itself define the situation, the 'facts of the situation' are an important con-
straint on the range of possible interpretations" (44).

Whereas the *logos* of nonexperts has its source in their lived and shared experience, the techno-scientific *logos* of institutions and institutional authorities
is considered to have its foundations in codified rules/procedures for argument, quantified data, and codified data-gathering practices. In the introduction to *Green Culture*, for example, Herndl and Brown explain, "The rhetorical power of this [scientific] discourse emerges from the rhetorical notion of
logos, the appeal to objective fact and reason. This is the discourse to which
the policymakers often turn to ground their arguments; technical data, and
expert testimony usually represent the basis of policy decisions, often at the
expense of other forms of rhetorical appeal" (11–12).

In their characterization of the *logos* of scientific policy argument, Herndl
and Brown recognize the oftentimes unbalanced power dynamic between in-
stitutional/techno-scientific *logos* and the *logos* of nonexpert citizens in which
the former is typically favored. This imbalance has been the focus of much
of rhetorical and sociological scholarship whose goal has been to restore
parity between the technical and nontechnical dimensions of policy argument. In some cases, this restoration has been approached by elevating lived-
experience to a place of prestige beside techno-scientific rationality. In other
cases, the definition of rationality has been expanded so that it includes a
broader spectrum of kinds of reasoning. In Walter Fisher's *Human Communication as Narration*, for example, he introduces the novel concept of "nar

rative rationality" that assumes parity between expert and nonexpert forms of argument. He explains, "The most fundamental difference in narrative rationality . . . [is] the presumption that no form of discourse is privileged over others because its form is predominantly argumentative. No matter how strictly the case is argued—scientifically, philosophically, or legally—it will always be a story, an interpretation of some aspect of the world that is historically and culturally grounded and shaped by human personality" (49).

In other cases, equity between expert and nonexpert *logos* is introduced by broadening the concept of expert knowledge such that it encompasses the input and knowledge of all participants in policy argument. Following up on Fischer's work, for example, William Kinsella argues that the *logos* that arises from the direct local experience of phenomena by nonexpert citizens should be seen as complementary to rather than distinct from the *logos* of techno-scientific experts:

> Enhancing public participation is important for preserving and strengthening democratic politics. . . . Nevertheless, the expert-public distinction continues to pose a practical and symbolic barrier to participatory decision making. Viewing expertise in broader terms as a public resource created through broad dialogue is one way of reducing that barrier.
>
> Ordinary citizens provide their own forms of expert knowledge when they contribute their local perspectives and values to policy decisions. Similarly, the contributions of specialists can be seen as one kind of local knowledge. (95)

The majority of rhetorical scholarship on the *logos* of expert and nonexpert participants in policy argument is devoted either to characterizing the *logos* of nonexperts or considering how to negotiate its differences with the dominant *logos* of science and the state. Very little attention, however, has been devoted to investigating whether and in what ways expert facts and methods might be productively integrated by nonexperts with their *logos* of lived experience and, thereby, influence their capacity to engage in policy argument. An investigation of the literature of rhetoric and communication studies[9] identified only a single case in which a rhetorical scholar had taken this issue up as the main focus of investigation. In "Science and Public Participation," Danielle Endres explores the role of techno-scientific *logos* in nonexpert public argument, concluding that there are three ways in which scientific expertise might be integrated into the public argument of nonexperts to support their case. These include "1) using scientific data produced by credentialed scientists to support a claim . . . ; 2) identifying flaws in the scientific method in order to challenge particular scientific findings; and 3) using one's own scientific data to make a claim" (55). In order to support these characteriza-

tions, Endres examines the public responses to the Yucca Mountain reposi-
tory that commented on or incorporated scientific claims. This analysis, however, falls short of identifying how scientific *logos* influences the argument
of nonexperts. In the paper's discussion of the public's efforts to critique the
scientific method, for example, the instances cited turn out to be informational inquiries by citizens about the scientific methods used for investigating risk rather than challenges to the *logos* of the scientific argument based
on adopted principles of scientific method (61–66). More relevant to the subject of citizen science and policy argument, Endres argues that the nonexpert
public can make claims using data from their own scientific research. However, she provides no examples or analysis of nonexpert argument based on
original citizen science to support her claim. As a consequence, this prom-
ising investigation offers suggestions but insufficient empirical illustration
of how techno-scientific *logos* might productively inform public argument.

Because citizen science involves the direct engagement of nonexpert members of the public with science and scientists, it provides fertile ground for
identifying concrete instances where techno-scientific *logos* might integrate
with and positively inform the *logos* of lay publics. This potential is evidenced
in a newly emergent scholarly discussion in the sociological literature[10] dealing with the topic. The first, and currently only, article in this literature is
Gwen Ottinger's "Buckets of Resistance." In her article, Ottinger uses the case
of citizen air quality monitoring in Louisiana to explore in some detail the
role of techno-scientific *logos* in political efforts to achieve environmental justice for residents living near an oil refinery. As a consequence of her investigation, Ottinger contends that the residents' citizen-science activities—using
specially equipped buckets to collect air samples in their neighborhoods—
resulted in an integration of techno-scientific and nonexpert *logos*. This integration, she claims, expanded the common ground of argument and evidence between the parties involved in policy debate[11] pressuring the EPA to
undertake their own official measurements of air quality in the neighborhood (246–47).

Despite citizen science's capacity to create new common ground of argument between experts and nonexperts, Ottinger also suggests that the case
illuminates the differences and inequalities between nonexpert and technical
logos in policy debate. She cites, for example, the conflicts between scientifically sanctioned measuring practices and those practices preferred by activist
groups and citizens. For EPA and industry scientists, air quality assessments
could only be considered legitimate if the air samples were taken over regular
intervals and then averaged. For citizens and their environmental justice advocates, however, measurements taken when the residents experienced negative physical symptoms were more meaningful because they supported their
goal of illustrating how bad the air quality could be in these communities.

By showing the potential of citizen science to create common ground while at the same time illuminating differences in the *logos* of expert and nonexpert arguers, Ottinger hopes that activists might better understand how to level the playing field in policy debates between expert and nonexpert arguers: "Perhaps the most surprising insight offered here is that the possibility exists for citizen scientists to use standards to their advantage, to exploit standards' boundary-bridging aspects to gain access to expert-dominated arenas. The challenge for citizen science, then, lies in making strategic use of standards—deciding in particular, which parts of the boundary to bridge and which detachments of border police to leave unchallenged" (265).

While Ottinger's work offers a glimpse of the importance of citizen science as a site for exploring an integrated *logos* and its value as a bridge between expert and nonexpert argument, her exploration is limited in a number of important ways. Though Ottinger shows that citizen-science activity can influence policy outcomes, there is very little evidence in her work of *how* it transforms actual policy argument. She explains for example that "by comparing peak data directly to regulatory standards . . . activists assert (albeit implicitly) that peak concentrations experienced in fenceline communities are consequential for long term community health" (257). Conspicuously missing from her assessment here, however, is any direct evidence from actual public discourse of this debate that would encourage her readers to accept this conclusion. Without corroborating evidence from the discourse of policy argument, Ottinger requires her audience to accept her conclusions on the basis of either the testimony of participants with whom Ottinger conducted interviews or on Ottinger's own recollections of the policy debate from her embedded experience with environmental justice organizations.[12] Although these represent important sources for understanding policy argument, they offer only general descriptions or recollections of the arguments made in policy deliberations. Without evidence from the actual discourse, however, it is difficult to ascertain in what ways and to what extent citizen science shaped the argument of the citizens who were involved or whether these nonexperts were even participants in the public debate over air quality.

In addition to providing limited primary evidence from discourse, there is very little, if any, discussion in Ottinger's work about the role that the context of argument might have played in influencing argument outcomes. Did social, political, material, or economic forces play a key role in compelling the EPA to do their own air quality testing in response to citizen science? Given the complexity of most policy argumentation, it is unlikely that a single factor, like a citizen-science project, would be sufficient to promote policy action. If we accept that policy arguments are complex and are often driven by a variety of factors, then it follows that it would be important to understand the features of context and their interaction with citizen science to establish the degree to which and the conditions under which citizen science might or might not impact policy outcomes.

The methods for analysis employed in this chapter endeavor to emulate the strengths and address the shortcomings in Ottinger's assessment of citizen science in policy debate. Like Ottinger's work, this analysis draws on interviews with participants in the policy debate to establish the general character of citizen science and its influences on policy outcomes. However, it also includes detailed rhetorical assessments of the discourse generated from both the oral and textual arguments in the policy debate. By assessing, for example, the arguments made by citizen scientists in a public meeting with local government representatives, it is possible to examine exactly how technical and nontechnical *logos* interact to support the policy arguments of citizen scientists.

In addition to providing concrete evidence of how participation in citizen-science activities might influence the policy arguments of nonexpert arguers, this analysis also explores how social, political, and argumentative contexts shape the policy arguments and outcomes of expert institutional arguers. It considers, for example, how plans for the redevelopment of Pepys Estate/Deptford influence the borough's responses to the conclusions advanced by the citizen scientists. It also explores how the borough uses the voices and conclusions of the citizen scientists to support their own arguments about the best policy solution to the problem of noise in Pepys Estate. By employing evidence from the actual discourse of policy arguments and details about the context in which they occur, this chapter offers both an illustration of how citizen science integrates expert and nonexpert *logoi* and how attention to context reveals the complexity of the role of citizen science in shaping policy arguments and outcomes. In doing so it moves beyond traditional rhetorical investigations of deliberative argument that consider expert and nonexpert argument as oppositional or, at best, incommensurable ways of arguing. It also presses further than the current sociological investigation of citizen science by offering direct evidence from discourse and argument of the impact of citizen science on policy argument and discussions of how contextual factors might interact with citizen-science efforts to influence policy outcomes.

Citizen Science on the Pepys Estate: A Tale of Two *Logoi*

The citizen-science project at Pepys Estate provides an ideal case study for investigating the central question of this chapter—To what extent and in what ways might citizen science influence public policy arguments and outcomes?—because community members gathered information about problems in their local environment through citizen science and used this information to create and present arguments directly to representatives from their local bor-

ough. The citizen science project at Pepys Estate was initiated in the summer/
fall of 2007 when Lewisham borough councilwoman Heidi Alexander was
made aware, through her connections with the nonprofit London Sustainability Exchange,[13] of the Mapping Change for Sustainable Communities project. She approached Muki Haklay about the project arguing that Deptford/
Pepys Estate was an ideal site for citizen-science mapping because of the
presence of a scrapyard that residents had complained for a number of years
was a noise nuisance ("Citizen Science" 1). The scrapyard was one of several
automotive-related businesses forming the Victoria Wharf Industrial Estate,
an island of industry in the center of Pepys Estate/Deptford surrounded by
residential high-rises and adjacent to a primary and nursery school. In response to the councilwoman's request, Haklay and his colleagues set up a
workshop on mapping on November 26, 2007, where they explained the mapping project to participating community members and visited the area around
which they would plan their mapping (Haklay, interview).

As a consequence of this meeting, four members[14] of the local community
volunteered to take sound readings. With help and guidance from London
21's Coleen Whitaker, residents measured noise levels over seven weeks from
January to February 2008 in a variety of locations around the industrial estate extending out to a maximum of 350 meters (.2 miles) ("'Citizen Science'
Takes Off" 2; Haklay, Francis and Whittaker, 28). This sustained program of
measurement yielded 385 sound measurements[15] and qualitative comments
on the sound being measured. Armed with their results, the citizen scientists and their academic colleagues approached the borough at the end of
March or beginning of April 2008 about setting up a public meeting to discuss the noise pollution issues in Pepys Estate. The borough agreed to meet
but requested six weeks to prepare their response to the data. At the end of
the period, a public meeting took place on the evening of May 15, 2008, that
involved the citizen scientists, representatives from the borough, members
of the neighborhood, and representatives of the Environmental Agency.[16] Almost all of the citizen scientists spoke[17] at the meeting (they are referred to
as "community ambassadors" in the audio recording) as well as the members
of the university and nonprofits who supported them, including Muki Haklay (UCL), Coleen Whitaker (London 21), and Gayle Burgess (London Sustainability Exchange). The borough was represented by Councilwoman and
Deputy Mayor Heidi Alexander and two borough pollution control officers,
Anton Murphy and Chris Harris. In addition, a regional representative of
the Environmental Agency, Matthew Wilson, was also in attendance. The remaining participants were local residents, young and old, interested in finding out about the noise pollution in their neighborhood and in voicing their
opinions about and solutions to the problem.

During the May 15 meeting, the policy arguments of the citizen scientists who participated followed an order typical of this genre of argument. As Fahnestock and Secor explain, policy arguments usually begin by arguing that a problem exists and then moves up the stases to address the causes and finally the solution to the problem (Fahnestock and Secor 292–97). In their presentations the citizen scientists made arguments about the existence, cause, and consequence of the problem, while the presentations by the representatives of the borough that followed offered arguments for what should be done about it. The first part of this analysis examines the presentations made by the three citizen scientists. The contents of these presentations— drawn from an audio recording of the meeting and the slides accompanying these presentations—reveal three prevailing arguments: 1) high noise levels in Pepys Estate exist, 2) the scrapyard is a significant source of noise, and 3) high noise levels have negative health consequences. By examining these arguments in detail, this assessment offers some answers to the inquiry, To what extent and in what ways can citizen science influence policy debate? It suggests citizen-science arguers integrated techno-scientific *logos* and nonexpert *logos* in their argument creating a hybrid *logos*.

This hybridized *logos* is illustrated clearly by the manner in which quantitative data about noise levels interacted with evaluative arguments about residents' lived experience with noise. As part of their experience in the sound-mapping project, the community ambassadors learned about the scientific method for quantifying noise levels using the measurement of decibels (dB) averaged over multiple recordings at a given time and on a given day ("Noise Mapping Toolkit" 1; Whitaker 2). In addition, they became familiar with international standards for acceptable sound levels and the possible negative consequences of exceeding those levels. Evidence that these expert methods and standards influenced the *logos* of their arguments about noise appears throughout their presentations: first in their argument about the existence of excessive noise levels and later in their conclusions about the possible consequences of excessive noise.

Making a Case for the Existence of the Problem

In the presentation of the first citizen scientist, qualitative and quantitative evidence intertwine to make the case that high noise levels exist on Pepys Estate. That the magnitude of noise is the topic of the first speaker's presentation is signaled at the very outset when she states, "We found that the noises in Pepys Estate in certain areas are very, very loud." Following this statement of the problem she offers the audience the quantitative data that proves its existence. Her quantitative argument begins by establishing the benchmarks for different levels of noise with the help of a technical table downloaded from the Internet in which quantitative decibel values are paired with descriptions of noise events typical for those values.[18] She explains using the chart, for instance, that street noise from traffic registers at 80 dBA.[19] In addition to using official scales to pair noise levels with noise events, she also uses benchmarks from institutional sources to indicate the normal or acceptable level for noise. In particular she draws on the noise thresholds set by the World Health Organization (WHO) in *Guidelines for Community Noise* (1999) in which they explain that "to protect the majority of people from being moderately annoyed during the daytime, the outdoor sound pressure should not exceed 50 dB LAeq"[20] (WHO, *Guidelines* 61).

Once the quantitative technical standards for defining acceptable or moderate noise levels are established, the citizen-scientist presenter turns the audience's attention to the data gathered by the sound-mapping project to make the case that elevated or unacceptable levels of noise exist in Pepys Estate. She begins her argument by describing extreme noise events. In her descriptions of these events, she integrates quantitative measures of the magnitude of sound with qualitative descriptions of the noise event. In her initial example, for instance, she relates her experience with measuring the noise levels produced by lorries (trucks) going to the scrapyard. She explains, "Sometimes when big lorries, huge lorries . . . were getting out or parking next to the scrapyard [the sound level reading] was going to 82 [dBA]. 82 is considerable between very, very noise[y] and noise[y]." In her second example, she describes the experience of another participant who measured the noise levels inside his apartment: "[D[21]] lives in Eddystone Tower. He collected sounds inside of his flat and it sometimes was 82. That is unbearable. If you leave open the window a little bit because it's summer, this would be too much for him."

In these two instances, we see quantitative and qualitative evidence working in tandem to amplify the argument that life in Pepys Estate can be extremely noisy. In this collaboration, quantitative evidence plays an important role, because it provides a criterion for judging the extremeness of these noise events. To say that truck traffic or your apartment is extremely noisy can be met with skepticism ("Is it really louder than anyplace else?") unless there is some comparative scale on which noise can be judged. In her argument, the first citizen-scientist presenter attempts to address this kind of challenge by using quantitative sound values whose magnitudes are contextualized with the help of the noise chart, which she introduced immediately before her anecdotes of extreme noise events. Without the quantitative decibel levels and the noise chart, it would have been difficult for the audience to judge the magnitude of these noise events. However, without the qualitative anecdotes the numbers themselves would likely have been much less impactful for the audience. The description of D's apartment, for instance, puts a human face on the experience of noise and its effect on individuals. It would have sufficed to say that a flat in Eddystone Tower had readings of 82 dBA to show that the dwelling spaces near the industrial estate were extremely noisy according to the noise chart. However, to imagine the life of a resident not able to open their windows on a hot summer day, constantly assaulted by extremely high levels of noise in their own home with no respite highlights for the audience that noise can have a very real impact on the life and livelihood of actual people. By blending quantitative with qualitative evidence, the speaker both deflects charges of subjectivism and establishes in a relatable way the existence of a problem.

Whereas the first presenter blends quantitative and qualitative evidence to make the case for the magnitude of the noise endured by residents, the second integrates these different forms of *logos* to argue for the frequency with which high noise levels occur. Making the case for the frequency of the problem is extremely important for the citizen scientists' case, because the presence of high noise levels on occasion would likely not be sufficient to compel the borough to act on the problem. To make a compelling case for the existence of a problem, they needed to prove that loud noise is a persistent part of their environment. To make this case, the second citizen scientist relies on the data, both quantitative and qualitative, which had been amassed from doing sound measurements. These data paint a statistical picture of a community living in an environment with persistent, unacceptable levels of noise. Out of the total number of noise measurements taken during the seven-week period (385 measurements), 264 readings (69%) had an average of over 60 dBA and 199 (52%) had readings of more than 65 dBA: respectively 10 to 15 dBA higher than the WHO's benchmark for acceptable noise levels. They also show that even evening (7 P.M.–11 P.M.) and night-time (11 P.M.–7 A.M.) noise levels could exceed WHO thresholds. In the evening, 70.3% of the measured noise levels (26 out of 37 readings) were above 60 dBA. At night, 44.8%, or almost half of the readings (13 out of 29), exceeded the 60 dBA threshold.

Although these numbers paint a compelling picture of a community existing in a soundscape persistently above acceptable noise thresholds, the significance of these data receive very scant attention until the presentation of the second citizen scientist. In her presentation, she draws the audience's attention to the frequency of sound events using a series of bar graphs that integrate the quantitative and qualitative data of the study. In the first two graphs, the quantitative data of the sound-mapping survey is combined with and compared to standardized noise-level thresholds. The first graph reacquaints readers with the relationship between qualitative sound descriptions (quiet, audible, loud, etc.) and the quantitative noise thresholds that correspond to them. On the graph, the qualitative descriptors are presented on the *x*-axis and the quantitative noise levels on the *y*-axis. The second bar graph[22] is similar to the first because it too presents qualitative noise levels on the *x*-axis. However, along the *y*-axis the quantitative noise thresholds are replaced with the number of qualitative assessment made by the citizen scientists for each noise level category (quiet, audible, loud, etc.). The bars on the second graph, therefore, allow the audience to visually compare the number of responses for each qualitative category. Based on this comparison, it is clear that the three categories with the most responses were "loud," "very loud," and "extremely loud," in descending order.

Though these graphs would likely have sufficed to make the qualitative case that the majority of respondents thought that sound levels were above accepted thresholds for noise, the second presenter also wanted to reinforce the point quantitatively by reviewing just how many responses were above threshold. Toward this end she makes an argument from accumulation by orally adding up the number of responses in each category: "We did 385 readings, yeah, and then if you add what is loud, plus very loud, and plus extremely loud. . . . If you sum over 120 readings [in the category of loud] plus over eight[y] [in the category of very loud], it's 200. Plus over more-or-less sixty [in the category of extremely loud]. If you sum this. . . . We got it very loud, loud, extremely loud 260 readings. It's too much. It's very, very annoying."

In this representation of the data, we again witness the mutual reinforcement of quantitative and qualitative argument. Here the number of readings are stated and added up culminating in a qualitative/quantitative conclusion that in the majority of the instances in which the citizen scientists gathered data the noise was loud, very loud, or extremely loud. This combination of quantitative and qualitative argument seems essential for establishing just how frequently these noise experiences occurred. Without quantitative evidence suggesting that noise experiences were above threshold 260 out of 385 times, the arguer would have had to rely on their subjective judgment that it was noisy "a lot" or "very often." This perspective, while representative of their experience, would not have been defensible within the context of a formal policy debate. Conversely, had the argument been made strictly using quantified figures—readings were over 60 dBA 120 times, over 70 dBA 80 times, etc.—the human lived experience of the sound would have been difficult to connect with, offering a less compelling exigence for immediate action. By using the qualitative descriptors "loud," "very loud," and "extremely loud," the arguer is able to offer the audience a relatable sense of the suffering of the residents and why action needs to be taken to end this suffering.

Making the Case for Cause

In the arguments establishing the existence of the problem, quantitative data and scientific criteria for judging noise levels were combined with the qualitative experiences of the citizen scientists. As the arguments moved from fact to cause, however, quantitative data played a less substantial role in the technical component of the argument. Instead, maps of the sound levels and locations of sound readings created by the global information systems experts at University College London filled the role of technical *logos*. Following the discussion of the frequency and quality of the noise described above, the second citizen scientist who had participated in data collection raises the question, "Where does the noise come from?" She begins answering this question quantitatively and visually with a bar graph showing the responses by the participating citizen scientists to the question, "What is the loudest sound you hear?" Of the eight categories of sound sources,[23] the two most identified sources of noise were "scrapyard" and "traffic" with traffic having slightly more identifications[24] than scrapyard. In her oral presentation, the second citizen scientist emphasizes these findings and makes the case that noise in a number of the categories, including traffic, can be attributed to the scrapyard: "Where came the noise? . . . Most of them come from the scrapyard and the traffic. If you see, if you think that the scrapyard [is responsible for] . . . all these things in this page that you have here [referring to other categories of noise sources on the bar graph] is responsible about the traffic and also is responsible about the lorry, then you can see that if a noise came from the scrapyard its very, very . . . it's too much."

By combining the traffic, lorry, and scrapyard, the second citizen scientist reinterprets for the audience the initial categories of causation. Instead of traffic, for example, being considered a primary cause of noise it becomes a secondary one whose noise must be attributed, at least in part, to the scrapyard, because presumably without the scrapyard there would be less traffic in the area and, therefore, less noise. This reasoning is also applied to the lorries (trucks) that are a constant presence around the scrapyard and can be directly linked to its operations. The qualitative responses about the sources of the noise from the sound-mapping survey, therefore, helped establish causation by providing a precise account of what the sound measurers identified as the loudest sound on the soundscape. Additionally, the sound survey provided the citizen-science measurers with opportunities to consider the relationships between the different sources of noise and develop arguments about additional causal relationships between the scrapyard and the truck/automobile traffic in the area, though these relationships are speculative rather than data driven.

In order to reinforce the initial argument that the scrapyard was the main cause of noise, the community ambassador turns the audience's attention to a map that indicates where readings were taken in which the scrapyard was identified as the loudest source of noise (figure 11). On the map each reading that fits this criterion is marked with an orange dot while readings from other sources are left white. At first sight the thick clusters of orange dots on Oxestalls Road and Grove Street draw the viewers' attention, which is not particularly surprising given that they represent sound readings taken in the

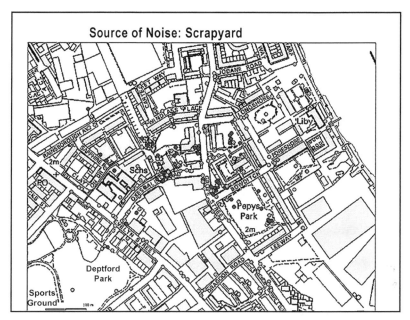

Figure 11. Places where the scrapyard was identified as the loudest noise on the acoustical landscape by Pepys Estate citizen scientists.

immediate vicinity of the scrapyard where we would expect it to be the loud-est source of noise. Interestingly, the speaker passes over these readings giv-ing them at most[25] a very perfunctory reference when she states, "This side comes from [the] scrapyard." Instead, she focuses her discussion on identi-fications of the scrapyard as the loudest source of noise that were taken the furthest away. She begins with the most remote identification that was taken by the Thames River: "If you see by the river, you can hear the scrapyard." She then moves to the third farthest identification in Deptford Park: "If you see here in Deptford Park . . . the loudest noise was from [the] scrapyard." Finally, she points to a cluster of identifications, some of which are remote and others of which are in close proximity to the scrapyard in the area of Pepys Park: "Here [in] Pepys Park and all around here the loudest [cause of sound] was [the] scrapyard, yeah."

By using the map of recoding locations where the loudest source of noise was identified as the scrapyard and by focusing on the most remote instances of these identifications, the citizen-science arguer is able to strengthen the case that the scrapyard is a significant cause of noise pollution in Pepys Estate. Whereas the quantitative/visual bar graph in the first part of her presentation is used to make a case about how frequently the scrapyard is mentioned as a source of noise, the map advances the argument that the scrapyard noise has a broad rather than localized impact on the community. This argument for the geographic scope of the scrapyard's causal influence cannot be meaningfully made by the quantitative tallies of the number of times it was identified as the primary source of noise. This is particularly true when we recall that many of the readings were taken in close proximity to the scrapyard where it seems unsurprising that the scrapyard would be the main source of noise. Making the case that the scrapyard is a pervasive cause of noise across the area requires mapping the data and highlighting the remote locations where the scrapyard was identified as the primary source of noise pollution.

As was the case with the first citizen scientist's presentation, we see also in the second's the integrated relationship between the nonexpert *logos* of lived experience and expert *logos* of techno-scientific assessment. Without the technical affordances for GIS mapping and the support from Haklay and his team in recording the precise locations of qualitative observations and mapping them, it would be difficult for the community ambassadors to make a compelling case to nonresidents that the scrapyard was the most dominant feature of the soundscape and that its noise had a community-wide impact. This illustrates the impact of citizen science supported and inspired by digital technologies on the arguments of the community ambassadors. In the absence of a community interpreter, however, there is no guidance into the meaningfulness of the map. It is only when the map is presented within the context of the lived experience of the community and their problems that the significance of the data and their visualization is revealed. In other words, without the second presenter's emphasis on the identifications of the scrapyard as a source of noise on the peripheries of the map, the audience would be unaware that the map's significance for this community member is how broadly the scrapyard was experienced as a source of noise across the neighborhood.

Making the Case for Consequences

Whereas the first two citizen scientists focused their presentations on making a case for the existence and cause of noise pollution, the final presenter examines the possible consequences persistent high levels of noise might have for the health of community residents. By examining the negative consequences of noise levels, the arguer highlights the potential risks for residents and, in so doing, attempts to generate emotional responses from different segments of the audience. From the borough representatives in the audience, the discussion of consequences is an attempt to elicit feelings of concern about the welfare of their constituents. From the members of the community, it is a strategy to evoke anger about being subjected to these risks. In both cases these emotional appeals are directed toward either promoting policy action or bringing pressure to bear on those that have the power to take action. Even in this emotionally charged finale of the community members' presentations, the techno-scientific *logos* generated by citizen science plays a central role in argument.

The contents of the third presentation can be divided into two parts. In the first part, the citizen scientist presenter supplies the audience with scientific facts drawn from a variety of medical, institutional, and media sources[26] that enumerate the possible negative health consequences connected with persistent exposure to high noise levels. The slides in this part of the presentation include titles like "Stress and Noise," "Autonomic Nervous System," and "Noise Induced Illness." With the opening slides, he makes two basic points about humans' physiological relationship with noise: 1) that we cannot control our physical response to it and 2) that noise impacts our autonomic nervous system and all of the organs connected to it (heart, stomach, brain). These points emphasize both the victimized status of those exposed to high levels of environmental sound and the scope of its impact on human health.

Once the general character and scope of the problem have been explained, the presentation supplies specific examples of the kinds of negative health effects that have been scientifically associated with excessive noise environments. These include hearing impairment, heart attack, depression, anti-social behavior, and even miscarriages. To endow these health effects with authority and objectivity, examples of these phenomena are drawn from a variety of scientific, institutional, and media sources. The use of authoritative sources and the sense of objectivity that they bring to the presentation are important for the citizen scientist's argument, because they help him avoid the charge that the health risks he claims the members of the community could be experiencing are subjective and, therefore, only apparent. Further, it authoritatively establishes links between particular health effects and noise.

Whereas part one of the presentation connects noise to specific health risks using authoritative sources, the second part offers details from the citizen-science project and from sources connected with the project to reinforce the point that residents in Pepys Estate are living in an environment in which noise levels are far above average. This reintroduction of local facts about noise levels following the presentation of health impacts would likely encourage the audience to imagine the deleterious effects that their noise environment might be having on them. On the first slide in this section "Noise Exposure," the presenter offers the audience "samples indicative of average noise levels inside residential dwellings in Eddystone Tower." Unlike quantitative argument in the previous presentations in which the citizen-science participants offered an average of many readings across the estate, here the reader provides only a few data points taken inside of his own home. This use of personalized data seems aimed at making the health risks of noise more proximal to the audience. These risks are no longer just some problem studied under managed conditions in technical settings by scientists. Instead, they are problems that have been explored inside of the apartments in which members of the community and audience live. In addition, it is no longer just faceless subjects in studies who have been exposed to levels of noise which are considered unhealthy. Rather, it is the speaker who stands before the audience who is a resident of the borough, a member of the community, and a neighbor who has been victimized.

With the aid of data gathered through the citizen-science project, the speaker provides the audience with a detailed quantified account of the level of risk he faces in his everyday life. Using averages from two days of noise measurements in January,[27] the speaker calculates that the average noise level in his apartment is 78.55 dBA. He also explains that he had recorded much higher levels in his apartment: "In my home I have recorded 89 decibels, excessively loud. From my window, I have recorded 109 decibels." Following these measurements, the speaker identifies his experience with that of the other community members in the audience and makes the connection between their collective day-to-day experience with noise in their homes and the negative health effects of noise: "Just bear in mind that we have been exposed to this noise over considerable periods. Bear in mind again it reduces the quality of all our lives. It also causes stress in children. [It's a] major impact on pregnancy. It's the cause of miscarriage in fact. [It is the source of] increased likelihood of strokes and depression as well as heart disease." In these lines, the use of the pronoun "we" as well as the phrase "all our lives" expands the problem of noise pollution from something experienced by the speaker in his apartment to a phenomenon common in the homes of residents in the audience. Following this identification he enumerates the health effects associated with high noise levels, many of which had already been introduced by slides and discussion in the first part of his presentation. This creates a link between the clinically defined health risks of noise and the lived experience of residents in the audience with noise. The value of this bridging is that it promotes a sense of immediacy and, therefore, exigence to take action or to appeal to others to take action.

In fact in the final slide of his presentation, the third citizen scientist attempts to capitalize on the exigence he has created by making the case that action needs to be taken to solve the problem. He opens his oral presentation of the final slide by summarizing the facts established in his presentation: "Noise harms us. It's not in dispute. It's well recognized by the medical profession. It's well known to legislative bodies. . . . Noise in residential areas is very important." The speaker then suggests, albeit obliquely, what needs to be done about the noise. He explains not once but twice, "Negative effects can only be reduced or relieved by the removal of the noise or its lessening." Although the scrapyard is never mentioned explicitly in this call for action, in the context of the previous two presentations, it seems clear that the third citizen scientist is advocating for either the complete removal of the scrapyard or at the very least some action by the borough to limit the amount of noise it produces.

By examining in close detail the presentations of the citizen scientists at the community meeting, it is possible to illuminate the way in which citizen science influenced the policy argumentation of nonexpert arguers. This section's close analysis of the oral, textual, and visual components of the argument suggests that rather than being in competition or conflict, techno-scientific *logos* (quantitative data, official units and criteria for assessing sound, and expert cartographic representations of data) and nonexpert *logos* drawn from lived experience complement each other in the arguments of citizen scientists. This hybrid *logos* was not limited to specific parts of the citizen scientists' arguments but played a seminal role in establishing the existence of the problem, its cause, and its consequences. This collaboration suggests that technical *logos* need not be thought of as fundamentally alien or hostile to the argumentation of nonexpert arguers. In fact, quite the opposite seems to be true. Citizen science empowered community members to develop a *logos* that validated their lived experiences while at the same time allowing them to frame these experiences in technical language and methods familiar to the borough representatives in the audience that they hoped to persuade to act. The questions that remain, however, are, Did the hybrid *logos* actually have persuasive effects on policymakers? and Did persuasion ultimately lead to actions favorable to the residents? In order to answer these questions, it is necessary to examine how the borough responded to these arguments and what policy actions they might have taken to address the concerns of their constituents.

THE BOROUGH RESPONDS: PERSUASION, EXIGENCE, AND THE FATE OF CITIZEN SCIENCE IN POLICY ARGUMENT

Accounts of the outcome of the Pepys Estate sound-mapping project hailed it as an exemplary instance of how citizen science driven by technological exigencies could empower communities to incite policymakers to act in their interests. In an online press release in June of 2008, just months following the meeting between the borough representatives and the citizen-science project participants, London 21 announced that the borough had taken action to respond to the issues raised by the citizen scientists: "Lewisham council & the Environment Agency accept that there is a problem. After seeing the results of the [citizen-science] survey, the Agency has appointed an acoustic consultant to carry out a detailed analysis of noise in and from the scrapyard" ("'Citizen Science' Takes Off" 1). The Environmental Agency's acoustical survey of the scrapyard began in July of 2008 and continued until January of 2010. During that time, the agency also advised residents of Pepys Estate/ Deptford to register official complaints about the scrapyard whenever it became excessively noisy. On the basis of the complaints and the acoustical measurements, the scrapyard's license was revoked in October of 2009 (SIMPLIFi Solutions). Despite being unlicensed, the yard continued to operate generating more citizen complaints and prompting the Environmental Agency to do video surveillance of the site between March and July 2010 to document unlicensed activity (Lawton and Briscoe 13–15). Armed with evidence of continuing illegal activity, the Environmental Agency requested a formal interview with the scrapyard's owner in September 2010. After receiving no response, they moved to officially prosecute the operation. As of October 2011, the site was "not accept[ing] any new waste and the illegal activity stopped along with almost all of the complaints" (11). Finally, on 19 December 2011, the Woolwich Crown Court levied a fine in excess of £230,000 against the scrapyard for its illegal activities (SIMPLIFi Solutions). Considering these details of the reported outcomes of the case, it seems that the citizen-science digital mapping project had been instrumental in provoking policy action by bringing the problem of the scrapyard to the attention of the authorities and providing the necessary proof for the borough to take action against the scrapyard. Though there is a degree of validity in these assumptions, a closer assessment of the borough's response to the sound-mapping data and their strategy for solving the problem suggest that the role of citizen science in initiating and encouraging policy action was neither straightforward nor simple.

The Persuasive Limitations of Citizen Science in Policy Decision Making

One of the ways in which citizen science proved to be a less-than-straightforward path from citizen action to policy change is that the proof it provided of the noise disturbance did not encourage the borough to shut down the scrapyard. Instead, they opted to undertake additional acoustic surveys of the site. This choice of action suggests that though borough representatives found the citizen-science data compelling, it was not sufficiently convincing to be politically actionable. The borough's need to establish greater conviction is evident in the responses of its officials to the presentations of the citizen scientists at the May meeting. In his opening speech, for example, Lewisham pollution officer Chris Harris congratulates the residents for the persuasive quality of their work. Immediately following these comments, however, he explains that the next step in the policy process is to hire a professional firm to take sound measurements: "I just want to say straight away that I agree with the findings. I agree from the point of view that this is a disturbance. . . . Now the answer to that is that we need to look at the noise. . . . We are working with the environment agency and we are going to be appointing an acoustic consultant to carry out a sound survey. . . . The plan will be to identify the areas we can work on within the site to try and control [the noise]."

In the opening line of Harris's statement, he explains that he agrees with the findings of the citizen scientists and proposes to take action. However, this agreement is tempered. The borough will take action, because its constituents have proved that they are disturbed by the noise, but not because their data is sufficient proof that immediate political/legal action against the scrapyard can be reasonably taken. Further, the action he suggests, an acoustic survey, is not a solution to the problem of the scrapyard noise; rather, it is an effort to establish the existence of the problem, an action that would have been unnecessary if the citizen-science data alone was sufficiently convincing.

The position taken by Harris and other borough representatives toward the citizen science violated the expectations of some of the community members at the meeting about the power of citizen science to establish conviction in policymakers about the problem. This violation of the expectations is illustrated by audience responses to Harris's suggestion that the next step should be an acoustic assessment of the scrapyard. In the discussion period following his opening statement, one women exclaims in frustration, "The point is I think it [the scrapyard] should just be got rid of now. Not just waiting for some future time. . . . We have had surveys done. . . . We have been collecting information for years. So it feels like [with a new acoustical survey] it's going back to the beginning again." Here we see illustrated the frustration with the borough's perspective that the data already collected are insufficient to shut down the scrapyard. Harris and Deputy Mayor Heidi Alexander recognize this frustration in their responses and make an effort to explain, and thereby justify, the borough's alternative criteria for establishing conviction:

> Heidi Alexander: Whilst the data you [citizen scientists] have collected is one form of data, this specific data [from the acoustical survey] will be in a format which could enable me to take decisions as well about the level of nuisance. I think it might be worth you [referring to Chris Harris] explaining why you need that different kind of data.
> Chris Harris: What this data [from an acoustical survey] could do . . . is that if there are [any] particular works that are going to be carried out, certain [noisy] activities [at the scrapyard] . . . we would then be able to look at them and compare them and also base them on the background levels of activities . . . and they could then see the intrusiveness and the improvements from cutting . . . certain activities so it goes into a lot more detail in the activities that are happening there and it helps us to then build a picture and see how we can move forward and get levels down.

In her opening comment, Alexander reaffirms the existence of two different perspectives about establishing conviction and is careful to differentiate between data that can create conviction and, therefore, serve as a justification for action within a decision-making context and data that cannot. She implicitly places citizen-science data in the latter category when she identifies it as "one form of data" in contradistinction to data from an acoustical survey that would be in a format to enable her to "take decisions" about what actions to pursue against the scrapyard.

Though Alexander's distinction between types of data makes the point that the kind of data gathered by the citizen scientists cannot inform policy decision making, it doesn't explain exactly why the data gathered by an acoustical survey would. This explanation is left to Harris whose answer is essentially that an acoustical survey would be more precise in its capacity to pinpoint where the noise is coming from. Although the results of the citizen-science project are clear that the scrapyard is a significant source of noise, they do not indicate precisely which aspects of the scrapyard operations are causing the noise. For residents the exact sources of noise were immaterial to their data gathering. Their goal was to prove that the scrapyard was a substantial noise nuisance. In doing so they believed that this would be sufficient for the borough to take direct action against the scrapyard. What they didn't account for was that the borough might consider a range of options including not shutting down the scrapyard but requiring them to stop or to mitigate the noise from particular parts of their operation. In the political context, considering this option might be preferable, because it would achieve lower noise levels while at the same time avoid charges that the borough was targeting the scrapyard and using heavy-handed regulatory tactics against it. For citizens, however, anything short of shutting down the scrapyard was an unacceptable response to the noise levels they had documented. By examining the details of the borough's response to the citizen-science project, it is clear that the relationship between citizen science and policy action against the scrapyard was complicated. Though the citizen-science project compelled the borough to take action, the actions that they took to gather more detailed information about sound were not the sort of actions expected by community members.

Redevelopment: The Hidden Exigence of Citizen Science

In addition to the less-than-direct relationship between the conclusions drawn from the citizen-science project and the borough's decision to take action, there is also a not-so-straightforward connection between the exigence accepted by the community as the rationale for citizen science and the exigence driving the borough's support for it. One of the important details of this case to consider is that the initial request to involve the Mapping for Community Change project in Pepys Estate/Deptford was made by the London Sustainability Exchange at the behest of Heidi Alexander, the ward's council-woman and deputy mayor.[28] At face value, this request could be considered a response by Alexander to requests from her constituents to document the noise levels at the scrapyard so that policy action could be taken. However, considering, as we have just done, that the borough representatives believed that the data from the sound-mapping project was insufficient to take policy action this raises the question, Why didn't the borough just do a professional acoustical analysis in the first place? Expediency could be one answer. The low/no-cost citizen-science project could provide the borough with unrefined preliminary evidence that a noise nuisance existed, which could help them decide whether a more extensive and expensive acoustical assessment was warranted. If, however, we consider the broader socio-political context in which the citizen science was taking place, a second alternative emerges.

As I have already mentioned, the borough of Lewisham is located in the east and south region of London, a region that has been the focus in the last decade of a massive project of redevelopment. Because of Pepys Estate/Deptford's location abutting the Thames and its situation as one of the poorest areas in London, it offered an ideal target for redevelopment. The scope of the redevelopment program devised for this area is outlined in a policy paper *Regeneration in Deptford, London* (2008), which lists at least seven[29] redevelopment projects proposed for this area of Lewisham borough during the period of the citizen-science project (Potts 19–31). Of these seven, four[30] were proposed for North Deptford, the area covered by the Pepys Estate neighborhood. Of these four sites one redevelopment site in particular, the Wharves at Deptford, has significance for this investigation, because its location covered the Victoria Wharf Industrial Estate, which included the scrapyard. As the Councilor for Evelyn ward and a Cabinet Member for Regeneration, Heidi Alexander was deeply involved in plans to redevelop the Pepys Estate/North Deptford neighborhood. One of her duties in this capacity was to secure the consent for redevelopment not only from landowners/tenants but also from the local community. Considering her leadership role in redevelopment as well as her goal of gaining consent for it, her request for a citizen-science project could be construed as a strategy for building neighborhood support for the redevelopment of the scrapyard site.

Evidence of a connection between the citizen-science noise survey and the redevelopment of the site is present in Alexander's initial comments at the May 15 meeting. She states, "We [the borough council] need to look at what we can do to try and ameliorate [in the short term] the really negative impacts [of the scrapyard identified by the community sound-mapping project]. In the longer term . . . we are looking at redesignating that site as a site that would be appropriate for different mix of employment . . . and some new homes." In these lines, Alexander informs the audience that the council already has plans for the site that the scrapyard is located on. She offers these plans for a mixed employment and residential development—the Wharves at Deptford—as a potential long-term solution to the problem of noise identified by the citizen-science project. To reassure the audience that the work of the citizen scientists has not been in vain, Alexander suggests that their efforts to gather data about the noise at the scrapyard can help galvanize political action in favor of redevelopment. She also defends the borough's move to commission an acoustical survey by suggesting that it is a necessary stop-gap measure meant to find ways of lowering the noise levels for residents until the longer-term political solution of redevelopment can be worked out: "A scrapyard is not a good neighbor for people when you are living on an estate. . . . There is someone that is actually buying up a lot of those plots of land there and . . . [who has] come forward with a redevelopment proposal. Now we need to work on our planning policies within the council to say if this developer buys a lot of the land could we make the case for something like a compulsory purchase order. This isn't going to happen overnight. It's a longer term issue. So in the meantime getting this consultant, acoustic consultant, in so we can look at ways we can lessen the impact is a positive step for us to take. But in the longer term that is part of the discussion we need to have about, you know, how we can get rid of the scrapyard."

A key piece of evidence connecting the citizen-science project to specific political action by the borough is made by Alexander when she explains that the borough might be able to "make the case" for a "compulsory purchase order" presumably because the scrapyard is not a good neighbor because it poses a noise nuisance. In this way, the data that the citizen scientists had gathered that the scrapyard was a noise nuisance, though not immediately convincing and actionable, in fact plays an important role in the long term for making the political case for its eviction.

Evidence that the borough draws on the citizen science and scientists to make the case for redevelopment appears in the materials generated for/by the borough in support of the Wharves at Deptford project. In the community consultation report *The Wharves, Deptford Statement of Community Involvement*, for example, the noise nuisance created by the scrapyard is advanced as a major source of community support for the redevelopment. The negative impact of the scrapyard is referenced a total of twenty times[31] and the phrases "noise pollution" or "the scrapyard" frequently top the list of community concerns that encourage the redevelopment. In a section of the document labeled "Lots to Like," for instance, the opening line reads, "The end of the scrapyard and car breakers and their early site clearance remain one of the most liked aspects of the proposal" (Soundings 17). The fact that the scrapyard plays such a significant role as a popular exigence for redevelopment in the community consultation report raises questions about whether the citizen-science project may have had an influence over community sentiments. A closer examination of the contents of these reports and the context of its development supports the conclusion that there are connections between the two. One feature of the context that suggests a connection is the time of consultation. According to the report, consultations of the community began in December of 2008 or a little over half a year after the May meeting between the borough and the citizen scientists. Although close proximity of time between the two events is circumstantial evidence of a connection, it supports the possibility that residents being queried about the problems of the neighborhood might have known about the citizen-science findings and had them fresh in their memories when they were consulted about the project.

Perhaps more compelling evidence of the influence of the citizen-science project on the document's arguments are allusions in the text to complaints raised by the citizen scientists in the May 2008 meeting with the borough. In a section titled "Noise Pollution," for example, the authors of the document explain, "A current problem for the local community is the noise at the car breakers and the scrapyard on the Oxestalls Rd. site. Lewisham council completed a noise audit after countless complaints by residents. We were also told that lorries delivering goods to and from the site posed a problem in terms of noise" (Soundings 122). In these lines, the authors of the document reference the arguments of the citizen scientists twice. At the end of the second sentence, they refer to the noise audit that was done after "countless complaints by residents." Considering the borough's response at the May 15 meeting, we know that the complaints that moved the borough to take action were the ones made by the citizen scientists and supported by their activities. This presumption is strengthened by the final line of the quote that refers specifically to arguments from the second citizen scientist's presentation, which causally connected the noise of the trucks and the noise of the scrapyard.

Perhaps the most convincing evidence that citizen science and scientists influenced the populist argument for redevelopment in the community consultation report is that people who were involved in the project were also interviewed by the report's authors. In these interviews, objections to the scrapyard are raised in each case. Though not directly involved in gathering data, Lewis Herlitz was intimately connected to the sound-mapping project and its findings as the moderator of the May 15 meeting between the citizen scientists and the borough. As the Director of Pepys Community Forum, he was consulted a number of times about the redevelopment.[32] Summaries of these consultations appear in the community involvement document. In one consultation,[33] the issue of the scrapyard and its noise is identified as a main discussion point. In a section titled "The Scrapyard," the authors of the document report that Herlitz commented, "It's got to go at some point.

You won't be able to attract anyone to live or do business in a development with a scrapyard" (Soundings 224).

In addition to Herlitz's words of support for removing the scrapyard, there are also arguments by two of the four citizen scientists who were involved in gathering data for the sound-mapping project. Dalva James and James (Jamie) Davies are reported to have also made the scrapyard a focus of discussion in their consultations on the Evelyn Community Garden.[34] Their comments about the scrapyard are highlighted in a section titled "Concerns about the Scrapyard," in which the authors of the document report: "Both Dalva and Jamie uttered their concerns about the negative effects the scrapyard has on the nearby residential areas. Noise pollution is very bad in the neighborhood. . . . Jamie said the neighborhood has been pressing the Council on this issue for years. Both strongly believe that the scrapyard should go" (Soundings 243).

As textual evidence from the consultation document reveals, the conclusions and experiences of the citizen scientists play a persistent role in making the popular case against the scrapyard in the document. Interestingly, however, neither the citizen-science project nor the participation of these respondents in the project is mentioned in the document. This raises the question, Does it really matter whether the borough or the consultancy agency[35] working with it knows or discloses the fact that the community members it has interviewed have been engaged in citizen science? I would argue that it does. Without reporting that James, Davies, and Herlitz had been actively involved with a project to confirm high levels of noise around the scrapyard, *The Wharves, Deptford Statement of Community Involvement* gives uniformed readers the impression that the evidence of community discontent with the scrapyard represents simply the lived experiences of community members. This representation is an inappropriate appeal to what I describe in chapter two as nontechnical ethical *arête*. By allowing the audience to assume that these community documents are the unvarnished, unprepared, and unmotivated expressions of local sentiment, the borough and the firm that created them falsely represent them as a free expression of public sentiment. If the document had explicitly recognized that some of the community's comments and reported opinions about the scrapyard had been developed through an organized program of research initiated by the borough's Cabinet Member for Regeneration, the extent to which these comments and opinions could be considered the natural expression of popular discontent with the scrapyard would likely have been challenged on ethical grounds by opponents of the redevelopment.

In addition to the ethical problem of disclosure, the use of the conclusions and the voices of the citizen scientists in *The Wharves, Deptford Statement of Community Involvement* also raises the question, Are representations of the problem and the proposed solution to the problem in the document commensurate with the views of those involved with the citizen science and the community they represent? In this case, there is very little question that the borough, the citizen scientists, and the community members living near the scrapyard were in agreement that the business was a local nuisance and needed to be gotten rid of. This does not mean, however, that these perspectives represent the views of everyone living in the Pepys Estate neighborhood. Interestingly, *The Wharves, Deptford Statement of Community Involvement* includes reports that residents not living in the immediate vicinity of the scrapyard did not consider it a nuisance and were not persuaded that it represented an exigence for redevelopment: "We observed that people who didn't live in the immediate neighborhood to the car breakers or the scrapyard thought that the site was 'just fine' and should stay the way it was" (Soundings 167). These alternative viewpoints are dismissed by the authors of the document in another section by claiming that, "There is an overwhelming support for the demise of the scrapyard shared by almost everyone except a small minority who tend to be people not living in the vicinity of the site" (210). Though there is no compelling evidence suggesting that the citizen-science project might have been use to amplify the extent to which the scrapyard was considered an annoyance by the community, it does raise the possibility that it could be used to this effect.

Perhaps a more serious concern is that the citizen science might have been used to justify a solution that was not fully supported by the community. As evidence in the previous paragraphs suggests, the community consultation document, whose goal was to present arguments for and against the Wharves at Deptford project, offered the removal of the scrapyard as the most commonly advanced reason supporting the redevelopment plan. In addition, it was revealed that the conclusions of citizen science and the testimony of citizen scientists were used in the report to make the case for the problem of the scrapyard. Because the arguments identifying the scrapyard as a problem were central for a document making the case for a redevelopment project, citizen science and citizen scientists are also implicated in supporting redevelopment as a solution to the problem. This raises the question, Is this solution supported by the community whose views on the problem are being represented using the conclusions of citizen science and the voices of citizen scientists? To answer this question, it is important to examine what if any objections might have been made against redevelopment as a solution and whether any of these objections could be traced back to members of the community associated with the citizen-science project.

In the community consultation report, there were a number of objections to redevelopment. The most significant of these was that by removing the scrapyard and the other businesses in the Victoria Wharf Industrial Estate the council would be eliminating employment in one of London's most impoverished areas. These objections were made by a broad spectrum of the community including those closely connected to the citizen-science project. Though supporting the eviction of the scrapyard, Lewis Herlitz, for example, was also clearly concerned about the impact of the redevelopment on the employment in the neighborhood. He hints at this problem in a reported exchange about shopping in Pepys Estate/North Deptford: "The area is poor and all the money is spent elsewhere. This is also a reason why local businesses are dying. There needs to be something that 'feeds back' revenue into the area to keep local businesses alive" (Soundings 219–20). A more direct challenge to redevelopment as a solution to the problem of the scrapyard was made by Malcolm Cadman, chair of the Pepys Community Forum's board. In his capacity as chair, Cadman knew about the citizen-science project and even posted about it on his blog.[36] His objection to the borough's redevelopment solution to the scrapyard problem is made clear in a letter written as the head of the Tenant Actions Group to the borough. He writes,

> We [the Tenants Action Group] wish to strongly object to the planning proposal for . . . the Wharves Oxestalls Road, Deptford. . . . LB [London Borough] Lewisham policy. . . . As far [as] the local community understands is that this site should be kept as a working area. . . . The type of work carried out at present on the site is beneficial to the local community. . . . The only deep concern with the employment use is the operation by Metal Recycling—known locally as "the scrapyard"—owing to noise and pollution. Yet the Environmental Agency (EA) is taking legal action and measures to remedy the situation. . . . While some redesignation of the land as residential in inevitable . . . employment space must be factored into the equation. . . . We ask that this application be turned down and remain with the existing policy to keep as a working area. (Cadman to LB Lewisham Planning, June 28, 2010)

In his capacity as head of the local Tenants Action Group, Cadman expresses a collective disagreement between a subset of the community and the borough over the solution to the scrapyard noise problem. Although he accepts that the noise from the scrapyard is a deep concern to local residents, in agreement with the community consultation report, he rejects the argument made by the borough and the report's authors that redevelopment is the best solution to the problem. Ultimately, there is no way to objectively verify whether his perspective or the ones in the consultation document truly reflected the views of all or most local citizens. However, the existence of a substantive difference between the borough and members of the local community who knew about or were involved with the citizen-science project suggests that citizen science could have been used to advocate for a solution to the noise problem that neither the citizen scientists nor their community supported.

Conclusion

The goal of this chapter has been to pursue the question, To what extent and in what ways might citizen science influence public policy arguments and outcomes? To explore this question, I have examined how citizen-science activities—inspired by an interest in finding policy applications for digital technologies—shaped the arguments of participants in the Pepys Estate sound-mapping project and how these arguments were received and coopted by representatives of the borough in policy argument. By examining how citizen scientists in Pepys Estate made their arguments about the scrapyard, this analysis moved beyond the present scholarship in rhetoric and sociology to reveal that the arguments of citizen scientists integrated the nonexpert *logos* of lived experience with the expert *logos* of empirically measured and quantitatively expressed data. In combination these *logoi* helped community members construct arguments that, on the one hand, resonated with policymakers while, on the other, expressed their personal and emotional experiences with the noise from the scrapyard.

In addition to illustrating how participation in citizen science influenced the arguments of Pepys Estate residents, this chapter also examined whether and in what ways they influenced policy outcomes. Though the general facts of the case suggest that the arguments made by citizen scientists served as a catalyst for change and led to outcomes in line with the goals of the citizen scientists and the community, a more detailed assessment of the borough's responses and the context in which these responses were made revealed that the success of citizen science in informing the policy process was not as straightforward as the reported accounts suggested. By assessing the borough's response to the citizen-science project, it was evident that though the borough was persuaded to take action, their choice of action, to make their own measurements, showed that citizen science provided insufficient verification of the problem to support policy action against the scrapyard. In addition, closer attention to the discourse created by policymakers and the context of this discourse revealed that there were multiple exigencies for the citizen-science project. They also revealed that the conclusions drawn from citizen science and the voices of citizen scientists were used by policymakers to support arguments about redevelopment. Assessing these arguments in the context of alternative community viewpoints about the existence of and solutions to the problem, however, suggests that the manner in which the conclusions of citizen science and the testimony of citizen scientists were used may not have reflected the interests or perspectives of the community, which supported and undertook the citizen science.

This detailed rhetorical analysis of the content and context of argument reveals that the role or potential role of citizen science, driven by technical exigencies to influence policy arguments and outcomes, is complex and offers benefits and hazards. On the one hand, digitally supported citizen science seems to provide a middle-ground where lived experiences of the average citizen and the epistemological requirements of policy argument can intersect. Though this intersection may not ideally satisfy the needs of one side or the other, at the very least it promotes the recognition that both sides are deeply interconnected in and important to the policy-making process. On the other hand, with these benefits come the potential for misunderstanding between policymakers and their constituencies as well as the potential for the inappropriate use of citizen science as a stand-in for popular sentiment. Although citizen-science activities may lead to outcomes that align with citizens' interests, there should be awareness among nonexpert participants that neither the speed with which nor the manner in which these outcomes can be achieved might match their expectations. In the case of Pepys Estate/Deptford, for instance, citizen scientists and their supporters expected that their data would be sufficiently compelling to encourage the borough to act immediately against the scrapyard and became angry when it was clear that it would not. The root of this discontent seems to be a basic misunderstanding on both sides about the differences between the requirements for proof in different spheres of argumentation. If citizen science is to be a productive middle way for argumentation, expectations about the kind of persuasiveness that can be achieve by citizen science and the potential obstacles in moving from policy argument to policy action need to be clearly defined. These expectations should be identified in the early developmental stages of citizen science to avoid later conflict. Because academic researchers in fields like Geographic Information Systems are typically focused on technical challenges at these stages, their citizen-science projects would benefit from the presence of rhetoric and communication scholars who are well versed in the challenges of argument and communication between experts and laypersons.

In addition to the capacity for creating false expectations about the strength of argument and its outcomes, citizen science also has the potential to be exploited by both policymakers and lay publics to make the case that their perspectives on policy problems and solutions are in line with popular sentiment. In the Pepys Estate case, the argument that the community was in favor of the Deptford Wharves redevelopment project as a solution to the problem of the noisy scrapyard is, potentially, a specious *ad populum* argument. However, the fact that citizen science might be used in this way is not cause to abandon it as an inventional source for policy argumentation, though it does call for transparency and careful scrutiny on the part of participants in policy debates. Policy arguers using citizen-science conclusions need to disclose any involvement they might have in initiating citizen-science activities. In the case of Pepys Estate, knowledge that the deputy mayor had been involved in the project from the beginning would certainly have given some decision-makers pause and opponents the opportunity to raise questions about the extent to which the community perspectives described in the document might have been unduly influenced by the borough's support for redeveloping the site.

The use of citizen scientists as stand-ins for community perspectives reveals the potential for citizen science and its connections to institutional power to become "invisible" in policy argumentation as well as the capacity of rhetorical methods to make these connections "visible." The value of the rhetoric/communication scholar's efforts to promote social justice and policy action through citizen science lies precisely in their detailed attention to what happens to citizen science once it has entered the realm of public argument and begun to play a role in persuasion. Professionals from other scholarly communities like environmental justice and public participation geographic information systems (PPGIS)[37] are invested in promoting digitally supported citizen science as a tool for social justice and community action. However, their involvement in these projects is typically limited to the inventional stage of policy argument, helping researchers identify technological resources or methods to promote their goals, and to the initial step in policy deliberation, opening up dialogues between communities and policymakers.

Once the data is gathered and members of the community and policymakers have been brought into contact with one another, these programs typically fold up shop and move on to the next project. In the cycle of policy *argument*, however, an initial meeting between decision makers and a citizen-science group would be the first stage of engagement, but not necessarily the last. As the case of Pepys Estate suggests, this engagement can continue months and years after the initial meeting during which time citizen-science evidence and arguments can be coopted and transformed in the process of public deliberation. The value of the rhetorical analyst is that their work begins where the engagements by other fields typically end. Through sustained attention to cases like the one at Pepys Estate, they may be able to offer advice to researchers in environmental justice, PPGIS, and urban planning about how to talk to residents and policymakers about what to expect in their argumentative engagements with one another. They could also function as public watchdogs on projects to ensure that important elements of context don't become invisible in the policy-argument process and that the representation of community perspectives in policy documents aligns with their policy goals.

By joining the skills and methods of rhetoric and communication scholars to those of technical researchers already invested in community engagement through digitally supported citizen-science projects, the gains made in empowering communities through data collection, online representation, and in-person interactions with policymakers can be sustained beyond initial contact as citizen science moves from the field into the realm of public argument and actions.

Epilogue

One of the challenges of writing a book about citizen science is that every year, even every month, seems to bring new developments. On February 14, 2014, for example, the citizen-science website Zooniverse reached a milestone by enlisting its one millionth volunteer. (In the three months since this announcement, an additional 107,000 have signed up.) Since its inception as Galaxy Zoo seven years ago, it has grown to include thirty different projects from astronomy to zoology, and its participants have made contributions to the study of space, climate, and the humanities with the publication of more than fifty research papers ("One Million Volunteers," "Published Papers"). A similar expansion in citizen-science activities has occurred in the projects chronicled by this book. Safecast, for example, has currently collected over four million radiation measurements and has expanded its mission from gathering data on radiation to collecting information on air quality. The Safecast Air project, which the group launched in 2012 with support from a $400,000 Knight News Challenge grant, is currently prototyping air sensors and has plans to start placing them in neighborhoods in Detroit, Los Angeles, and Tokyo (LaFrance).

As scientists strive to extend the boundaries of knowledge and citizens take action to understand techno-scientific risks, digital-age citizen science provides them both with the material resources and the epistemological empowerment to pursue their goals. In this book I have endeavored to raise awareness about this emerging phenomenon and its importance for rhetorical scholars by examining 1) the influence of digital-age technology on argument and 2) the capacity of citizen science for shaping relationships between laypersons, science, scientists, and decision makers. With the case studies complete, I want to take the opportunity in this epilogue to revisit these general themes, consider their significance, and contemplate future avenues for exploration.

The Influence of Technology
on Communication and Argument

The influence of the Internet on argument has piqued the interests of researchers in rhetoric, particularly those exploring political messaging and rhetoric. Less attention, however, has been paid to how the spread of digital technologies has affected argument about techno-scientific issues. This book explores this largely uninvestigated topic by examining Safecast's efforts to measure and represent radiation risk in the months following the Fukushima accident. In chapter 2, I show how contextual factors like risk information, politics, and the emergence of digital technologies have shaped the visual representation of radiation risk by comparing risk visuals from the accidents at Three Mile Island, Chernobyl, and Fukushima. I also suggest that choices of visualization can be explained by variations in the goals and audiences of risk communicators. While Safecast created detailed visualizations to educate and inform Japanese citizens about their particular exposure to risk, sources in the mainstream media created general risk visualizations which reflected their goal of providing a broad overview of the accident to non-Japanese audiences.

As the first sustained effort to examine visual risk communication, this chapter breaks ground by highlighting previously unrecognized rhetorical dimensions of this mode of communication. It reveals how social, political, and technological context; conventions of representation; and audience shape how communicators choose to visualize risk. It also recognizes the possibility for competition between grassroots and institutional risk representations. By examining these competing representations it raises awareness of the limitations of institutional risk visualization strategies and considers the opportunities for developing more citizen-centered forms of visual risk communication.

Whereas chapter 2 investigates the influence of digital technologies on communication, chapter 3 examines their impact on argument. In particular, it explores whether, and if so how, digital technologies might help laypeople overcome the problem of expertise in public argument. The role of scientific expertise in argument has been an important subject for rhetoric and communication scholars who have investigated such diverse topics as the common strategies for appealing to expertise (Hartelius 2011), the collective ethos of expert groups (Keränen, *Scientific Characters* 2010), and the capacity of experts to speak on subjects beyond their expertise (Lyne and Howe 1990). They have also turned their attention to the problem of scientific expertise in public advocacy. In their work on AIDS activism, for example, Valeria Fabj and Matthew Sobnosky point out that cultivating expertise is essential for entering into public sphere deliberations: "It is at this point [when lay people feel free to speak publicly on AIDS] that people need to be informed and to learn the language of science so as to be good consumers of science and to enter the conversation themselves" (182). Despite a general interest in the rhetoric of expertise and specific attention to the problem of expertise in public participation, rhetoric scholars have only just begun to examine how digital technologies might help nonexperts cultivate expertise. With the aid of Collins and Evans's classifications of expertise and classical perspectives on *ethos*, I analyzed how Safecast represented its goals and practices in the media. My analysis showed that Safecast's increased participation in data-gathering activities, supported by their development of the bGeigie, correlated with an expansion in the group's available means of persuasion from strictly nontechnical ethical argument to both nontechnical and technical ethical appeals. This correlation suggests that the Internet and Internet-connectable devices have the power to erode the obstacles created by expertise in public sphere argument.

The case of Safecast also raised questions about whether models of expertise were adequate for classifying citizen-science activities. In chapter 3, I argue that Safecast's radiation data gathering did not fit neatly within Collins and Evans's category of "contributory expertise" because it was not driven by scientific exigencies. I proposed that in order for this discrepancy to be accounted for, "contributory expertise" should be divided into two sub-categories: "technical/informational" and "analytical" contributions. This distinction accounted for Safecast's development of "contributory expertise" by participating in disciplined and scientifically informed data-gathering practices. At the same time, it acknowledged that the contributions and exigencies of this participatory expertise fundamentally differed from the scientific ones recognized in Collins and Evans's model.

The challenge of situating Safecast's citizen science into Collins and Evans's model of experience and the reconfigurations required to make it fit raises questions about whether the book's other cases might pose similar problems. A brief assessment suggests that they do. Chapter 4, for instance, examines how climate change skeptic Anthony Watts gained access to the technical sphere debate over temperature measurement. Unlike the citizen science of Safecast, Watts's Surface Stations project was guided by scientific questions about the influence of site conditions on temperature measurement. Though researchers at the National Climatic Data Center (NCDC) rejected Watts's conclusions, they respected the integrity of his data and incorporated them into their own research to draw conclusions about siting biases. In this way the results of the citizen science became thoroughly imbricated in the scientific conversation over climate change. Surface Stations, therefore, fits more comfortably than Safecast into Collins and Evans's category of "contributory expertise"; however, this characterization is not without issue. Though Watts did participate in the technical sphere debate, he was not a climate scientist. Further, he was more heavily influenced by political exigencies than scientific ones in his pursuit of answers to the scientific research questions. On these grounds, it might be more reasonable to suggest that his expertise uncomfortably straddles the divide within "contributory expertise" between "analytical" and "technical/informational" expertise.

The final case study on the Pepys Estate sound-mapping project provides a further example of the limits of Collins and Evans's model for capturing the complexity of expertise developed by citizen scientists. Unlike Safecast, the citizen scientists of Pepys Estate were not independently motivated to develop the technologies or learn the data-gathering methods for assessing noise pollution. Instead, their engagement with expert knowledge was initiated from the top down, mediated by researchers at University College London (UCL). UCL researchers helped Pepys Estate residents use sound meters and digital maps to develop arguments to support their case for neighborhood noise pollution. Unlike the data gathered by the participants in the Surface Stations project, however, the information collected by the Pepys Estate citizen scientists could not stand alone in technical sphere argument. Instead, it was only sufficient to convince the borough to initiate an expert acoustical survey. These characteristics locate the citizen scientists in this project on the continuum of expertise somewhere between Collins and Evans's categories of "interactional expertise" and "contributory expertise." Because the citizen scientists found support for their risk claims through hands-on research using technologically disciplined data-gathering techniques, the project had some of the characteristics of "contributory expertise." However, it also had characteristics of "interactional expertise," because the citizen scientists' expert knowledge was gained through sustained interactions with UCL researchers, who initiated the project, collected and organized the technical knowledge, and trained residents in the methods of data gathering. This case is interesting because "interactional expertise" paved the way for "contributory expertise." It is also significant that the "contributory expertise" of the citizen scientists was guided by a socio-political exigency rather than a scientific one and that its contribution, though informational, was only preliminary and required more data gathering by experts for the problem to be certified as existent and actionable.

The challenges Collins and Evans's model faces trying to accommodate the kinds of expertise developed by citizen scientists in these three cases suggest that the model, while an excellent starting point for thinking about expertise, can be productively revised and extended. Most significantly the concept of "contribution" in the category of "contributory expertise" can be expanded to include not just the exigencies of scientific communities but social and political exigencies as well. This permits the model to accommodate what is apparently unique about citizen science: that it involves laypersons developing scientific expertise not to further scientific knowledge and practice but to engage in socially and politically motivated techno-scientific risk identification and argument. Because citizen science implicates science in social and political debate, it moves the conceptual center of Collins and Evans's model away from science and onto the boundary between the technical and public spheres. Given that science is frequently defined by its goal of creating new knowledge, this shift away from epistemology and toward social and political action suggests that the "science" in citizen science is expanding beyond its traditional borders to inhabit a more conceptually diverse and rhetorically rich territory. In addition to reorienting expertise to include nonscientific goals, citizen science also foregrounds the importance of practice in the development of expertise. In Collins and Evans's model, the division between the scientist and the nonscientist is predicated on the idea that practice is the exclusive domain of the scientific expert. The case studies in this book have shown, however, that with the growth of digital technologies, practices that were once the exclusive domain of scientists, such as gathering massive amounts of data on radiation, are now accessible to tech-savvy laypersons. This move toward practice by laypeople destabilizes Collins and Evans's division between scientists and nonscientists and reaffirms the importance for rhetorical scholars of studying scientific practice along with the social, argumentative, and linguistic dimensions of science. In combination, these changes in exigence and practice suggest that traditional perspectives on science on which Collins and Evans base their model are beginning to lose their capacity to faithfully reflect what it means to do science or be a scientist in the case of citizen science. As chapter 1 illustrated, these nontraditional scientific enterprises have been around for at least a century; however, the emergence of digital technologies has legitimized and popularized them increasing their capacity to complicate traditional perspectives on science.

In addition to illuminating the challenges created by citizen science for sociological models of expertise, the cases in this book also point out the shortcomings of current typologies of citizen science developed in scientific scholarship. Scientific classifications of citizen science, like the one found in the CAISE report, are limited in a number of important ways. First, they fail to recognize citizen-science projects that are initiated and developed by laypeople. This limitation is likely a consequence of the fact that scientific typologies of citizen science are created by and for scientists and assume a central role for them in developing citizen-science projects. Another shortcoming of scientific models is that distinctions between different kinds of citizen science don't account for the social or epistemological impacts that citizen science can have. For example, the difference between the lowest category of participation in the CAISE report's model, "contributory projects," and the next highest one, "collaborative projects," is that in the former category laypersons are strictly data gatherers for scientists whereas in the latter they help analyze data and disseminate results. Though disseminating results suggests the possibility for impact, assessing what the impact might be and using it as a feature of classification goes beyond the parameters of the CAISE model. The instances of citizen science examined in this book, however, suggest that the kind of impact a citizen-science project has distinguishes it from other kinds of projects. Close examination of the arguments in the Pepys Estate sound-mapping project, for example, showed that citizen science with mixed technical and nontechnical *logoi* was not particularly impactful in policy debate. Though it persuaded the borough to act, these actions were not in line with residents' expectations. In comparison the Surface Stations citizen project had a significant impact on public and technical sphere debate. It not only helped Watts and his perspectives get widespread media attention, but also gave him influence in the scientific debate as well. Without considering the impact of citizen-science activities, or what criteria might be used to assess them, scientific models fail to look beyond the immediate configuration of physical and intellectual labor in citizen science to consider how the products of these configurations shape knowledge, belief, and action in the world. By introducing impact as a category, we can have a richer understanding of who the target(s) for impact are, what kinds of possible outcomes citizen science might support, and what factors might influence the degree to which citizen science can be impactful.

Though this book's investigation of the influence of digital-age citizen science on argument and communication benefits our understanding of visual risk communication and the impact of digital technologies on expert argument, there are still important ways in which it might be expanded. A significant subject that was touched on but not fully elaborated in the book's third chapter was how digital technologies and technical lines of argument might impact the identities of lay arguers or the groups they form. In his research on AIDS advocacy, sociologist Steven Epstein suggests that such transformations can happen when nonexperts engage with technical subjects. He explains, "As activist leaders . . . become full-fledged experts, they have often tended to replicate the expert/lay division within the movement itself by constructing a divide between the 'lay expert' activists and the 'lay lay' activists" ("Construction" 429). Because the development of expertise is a significant subject in the study of citizen science, the phenomenon provides a fertile space for exploring how changes in expertise might influence group identity and dynamics. Future research could inquire, Could adopting expert modes of communication and argument precipitate an identity crisis within a grassroots citizen-science group? Could it create social or epistemological distance between a group and the lay constituencies they serve? Do grassroots groups facing these challenges develop adaptive strategies for communication and argument to respond to these changes? By pursuing these questions rhetoric and communication scholars could use citizen science as a space to explore the consequences of expert modes of argument on group identity and integrity as well as the role of nontechnical, rhetorical forms of argument in maintaining group cohesion and identity during transitions from nonexpert to expert forms of communication and argument.

Shaping Relationships between Laypersons, Science, Scientists, and Decision Makers

As scholars in rhetoric and communication studies have pointed out, the manner in which relationships between laypersons, science, scientists, and decision makers are configured and their capacity to be reconfigured can have significant consequences for interactions between the private, public, and technical spheres. In *Scientific Characters*, for example, Lisa Keränen argues, "Reconfigured relationships between science, its stakeholders, and publics . . . have the potential for promoting democratic engagement . . . by allowing members of various publics to deliberate with scientists and policymakers on matters of mutual interest" (165). Scientific researchers also recognize that relationships between laypersons and scientists can be shaped, and through this shaping have influence on the way the public perceives scientists and scientific perspectives. As Janis Dickinson and Rick Bonney of the Cornell lab of Ornithology have argued, one of the benefits of citizen science is that it can improve laypeople's understanding of science and promote their identification with scientists and scientific viewpoints. In their book *Citizen Science*, they explain, "Citizen science has the potential to build important bridges between professional scientists and the public with positive outcomes for both science and public scientific literacy" (10). Though scientists believe that citizen science can have these benefits, they have not yet systematically studied its efficacy. Dickinson and Bonney remark, "although evidence suggests that participants in some projects may begin to 'think scientifically,' citizen science has not been well studied for its potential to change peoples' perceptions of science and of themselves as scientists" (11). By adopting a rhetorical perspective that focuses on the social and political dimensions of citizen science, this book has examined whether, and if so how, citizen science might be influencing the relationships between laypeople, science, and scientists.

Toward this end chapter 4 explores the Surface Stations project that was a citizen-science collaboration between the climate change critic Anthony Watts and the climate scientist Roger Pielke Sr. A rhetorical assessment of this project showed that nonscientific factors could neutralize the benefits of even the most collaborative of citizen-science projects. An analysis of Watts's *Is the U.S. Surface Temperature Reliable?* suggests that his personal commitments to a critical position on climate change and pressures from audience expectations may have encouraged him to develop a historical narrative of the project that obscured Roger Pielke Sr.'s role as its intellectual inspiration. I also demonstrate that institutional values and goals could influence representations of citizen science. I give evidence, for example, that researchers at the NCDC had used value arguments to elevate their quantitative methods for assessing the biases of temperature measurement. These value arguments were meant to challenge the qualitative methods pursued by the Surface Stations project that the NCDC believed threatened to turn the public against their stance on anthropogenic climate change. This evaluation of the Surface Stations project illustrates that though citizen science promoted interaction between laypersons, science, and scientists these interactions were not themselves sufficient for establishing better relations between them.

Another proposed benefit of digital-age citizen science is that it provides nonexperts with the opportunity to more closely engage with technical information and expertise and thereby strengthen their ability to advocate for their interests in the public sphere. Gwen Ottinger recognizes this benefit in her paper on citizen science air quality monitoring where she writes, "By using real-time air monitors instead of sampling devices . . . activists could document both average chemical concentrations and peak exposures; the contrast between the two could form the basis for pointed attacks on the [EPA's] ambient air standards' implicit claim that only averages matter to health" ("Buckets" 266).

Like Ottinger's work, the final chapter of this book explored whether, and if so how, citizen science supported the policy goals of laypeople. By assessing the actual arguments made by the citizen scientists of Pepys Estate, I showed how participation in citizen science helped them develop technical *logos* to make their case. This *logos* was sufficiently compelling to persuade the borough to commission a professional technical assessment of noise levels to further explore the problem. However, it was insufficient to encourage local officials to pursue community-supported solutions, an outcome that was not fully anticipated by either the residents or the citizen scientists who had gathered the data. What neither foresaw was that though the borough could use the citizen-science data to get rid of the local scrapyard, it could also use it to argue for replacing the scrapyard with a housing redevelopment.

The existence of different exigencies for residents, scientists, and policy-makers in the Pepys Estate case suggest that citizen science does not always shape the relationships between laypersons, science, scientists, and decision makers in predictable ways. In fact, the existence of multiple agendas indicates that there can be competing goals and unforeseen outcomes when citizen science is used in policy advocacy. The complexity of the relationship between science and policy illustrated by this case means that assessing the outcomes of citizen-science projects will likely require inter-disciplinary expertise. Though scientists, for example, may be able to assess whether citizen scientists have gathered reliable data or developed a deeper appreciation of science, they may not be prepared to investigate the way in which socio-political contexts influence how citizen science gets used in the public sphere. In the final chapter, for example, I showed that researchers from University College London largely ended their involvement with the citizen-science project after the initial meeting between the borough and the residents of Pepys Estate. Without following the impact of citizen science through the stages of policy debate, these researchers missed the possibility for conflict between the borough and residents over the solution for redevelopment.

To appreciate the full complexity of the relationships generated by citizen science and their influence on outcomes, specialists from fields like sociology, policy analysis, communication studies, and rhetoric should also participate in research on citizen science. In my assessment of the Pepys Estate case, I have shown the benefits of using a rhetorical approach to assess the phenomenon. Unlike current efforts in other disciplinary fields, a rhetorical approach examines the actual discourse generated both orally and textually from the policy debate. This approach allows the voices of all sides in public deliberation to be assessed to create a detailed understanding of the dynamics between laypeople, science, scientists, and policymakers throughout the policy process. In addition, a rhetorical approach provides the contextual details necessary to create an informed interpretation of this evidence and these relationships. By adding historical details about regional redevelopment plans and the efforts of the borough councilwomen to actively recruit researchers to do citizen science in Pepys Estate, I was able to show that while citizen science could drive policy decisions, policy decisions could also drive citizen science.

Like the study of communication and argument, the exploration of how citizen science influences relationships between laypersons, science, scientists, and decision makers invites questions that require further investigation. Prominent among these is how citizen science can be planned to more effectively facilitate identification between scientists and laypeople. In my discussion of the Surface Stations project, I argued that the collaboration represented a missed opportunity for climate scientists to reframe the narratives created by climate skeptics. By deciding not to review the discourse and argument in Watts's report, Pielke Sr. gave up the opportunity to identify and address pejorative conservative representations of the US government and government scientists as uncritical supporters of climate change. Had he done so, he may have been able to offer an alternative narrative highlighting the critical stance some climate scientists had taken on the problems of temperature measurement. What this case suggests is that though citizen science does have the potential to shape the relationship between laypersons, science, and scientists, its potential can only be realized if the rhetorical dimensions of citizen science are taken seriously and addressed.

In my assessment, however, I neither asked nor pursued questions about how this might be accomplished. What sort of protocols would need to be created to ensure the coordinated and collaborative production of discourse about project goals, methods, and the roles of participants in data gathering and assessment? Who would be the audiences for these documents? What sort of accommodations would need to be made to meaningfully address them? Answering these questions will likely require rhetoric and communication scholars to leave the comfort of texts and seek out real citizen-science projects to embed themselves in. As embedded participants, they could use their knowledge of argument and communication strategies and tolerance for *dissoi logoi* to help both sides understand their exigencies, goals, and imagined audiences for the project. With this knowledge, they could guide participants in developing collaborative communications by illuminating for them how their strategic choices of language and argument advance their individual interests and perspectives as well as what alternative collaborative representations might be available. This kind of embedded and applied research would be valuable for rhetorical scholars, because it would help them understand the goals and strategies of the parties involved in cocreated citizen-science projects as well as the challenges of negotiating and crafting documents reflecting both the particular interests of the parties involved as well as the bipartisan spirit of the endeavor.

Conclusion

With this book, I have highlighted a few of the significant issues raised by citizen science in the hopes that these investigations might pave the way for a sustained study of the phenomenon. Because it bridges the gap between the technical and the public sphere, citizen science provides a space for investigating topics of significant interest to rhetoric including the role of visuals in risk communication, the contribution of digital technology to the development of expert argument, the challenge of creating identification between scientists and laypersons, and the interaction of techno-scientific *logos* and nonexpert *logos* in policy debate. With the continuing growth of digital technologies, the ever-increasing material needs of science, and the expanding use of science in public policy, digital-age citizen science is likely to have broader and more diverse applications and impacts in the coming decades. By working collaboratively, scholars in the humanities, sciences, and social sciences can help laypeople, scientists, and policymakers more productively collaborate through citizen science to develop solutions for the scientific and social problems of the twenty-first century.

Appendix A

Acoustical Assessment Form Used in Pepys Estate Noise Mapping

---

Pepys | Noise Mapping

Name: _____ **Date:** _____

All the data collected will be processed anonymously. Your name is only requested here so that we can contact you with any technical queries.

---

(4) Mark the location of this reading on the attached map with "X4"

Time	1st Sound Reading (dBA)	2nd Sound Reading (dBA)	3rd Sound Reading (dBA)

Please circle any words that describe the quality and intensity of the sound:

What is the loudest sound(s) you hear:

Silent	High pitched	Constant	Enjoyable
Extremely Quiet	Shrill	Repetitive	Relaxing
V. Quiet	Sharp	Intermittent	Comfortable
Quiet	Spikey	Abrupt	Acceptable
Audible	Hollow	Random	Annoying
Loud	Deep	Indistinct	Exhausting
V. Loud	Bassy		Disturbing
Extremely Loud	Low pitched		Threatening
Painful			Agonizing

Additional comments:

---

(5) Mark the location of this reading on the attached map with "X5"

Time	1st Sound Reading (dBA)	2nd Sound Reading (dBA)	3rd Sound Reading (dBA)

Please circle any words that describe the quality and intensity of the sound:

What is the loudest sound(s) you hear:

Silent	High pitched	Constant	Enjoyable
Extremely Quiet	Shrill	Repetitive	Relaxing
V. Quiet	Sharp	Intermittent	Comfortable
Quiet	Spikey	Abrupt	Acceptable
Audible	Hollow	Random	Annoying
Loud	Deep	Indistinct	Exhausting
V. Loud	Bassy		Disturbing
Extremely Loud	Low pitched		Threatening
Painful			Agonizing

Additional comments:

---

(6) Mark the location of this reading on the attached map with "X6"

Time	1st Sound Reading (dBA)	2nd Sound Reading (dBA)	3rd Sound Reading (dBA)

Please circle any words that describe the quality and intensity of the sound:

What is the loudest sound(s) you hear:

Silent	High pitched	Constant	Enjoyable
Extremely Quiet	Shrill	Repetitive	Relaxing
V. Quiet	Sharp	Intermittent	Comfortable
Quiet	Spikey	Abrupt	Acceptable
Audible	Hollow	Random	Annoying
Loud	Deep	Indistinct	Exhausting
V. Loud	Bassy		Disturbing
Extremely Loud	Low pitched		Threatening
Painful			Agonizing

Additional comments:

(Developed by: London 21, UCL and Christian Nold)

Appendix B

Sound Level Reference Chart Used
by Pepys Estate Citizen Scientists

TYPICAL SOUND LEVELS

Jet takeoff (200 feet)	120 dBA	
Construction site	110 dBA	*Intolerable*
Shout (5 feet)	100 dBA	
Heavy truck (50 feet)	90 dBA	*Very noisy*
Urban street	80 dBA	
Automobile interior	70 dBA	*Noisy*
Normal conversation (3 feet)	60 dBA	
Office, classroom	50 dBA	*Moderate*
Living room	40 dBA	
Bedroom at night	30 dBA	*Quiet*
Broadcast studio	20 dBA	
Rustling leaves	10 dBA	*Barely audible*
	0 dBA	

Source: http:/ / www.brcacoustics.com/noisedescriptors.html

Appendix C

Bar Graph of Qualitative Sound Descriptors
for Pepys Estate Sound Mapping

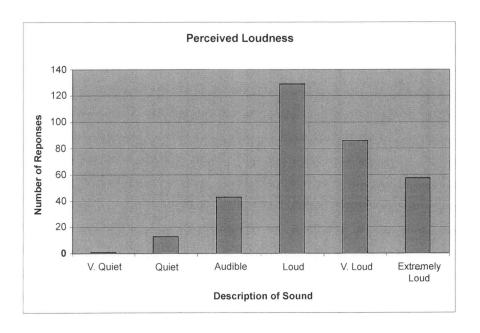

Appendix D

Bar Graph of Sources of Noise on Pepys Estate

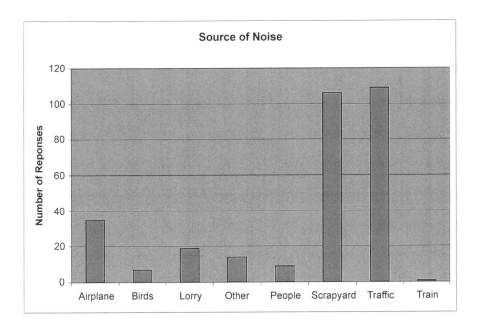

Notes

INTRODUCTION

1. Authors of modern discussions of citizen science point back to the use of citizens to gather data about astronomical phenomena in the nineteenth century and the Christmas Bird Count, which began in the early twentieth century, as past instances of citizen science. Chapter 1 deals with the history of citizen science in depth. See also Silvertown 467; Dickinson, Zuckerberg, and Bonter 150.

2. A search of the *Web of Knowledge* database shows that the first instances of the phrase "citizen science" in the database appears in 2004. I am assuming that Silvertown's calculations are based on the number of references between 2004 and 2009.

3. Phenology is the study of periodic events in plant and animal life cycles such as blooming or egg laying. See "Phenology."

4. The USHCN is a system of sensors used by the US government to calculate warming trends and design climate models.

5. My contribution in chapter 10 of *Science and the Internet* includes additional material complementing the investigation in chapter 2 of the visual representations of radiation risk.

6. *Rhetoric Society Quarterly, Rhetoric and Public Affairs, Environmental Communication, Technical Communication Quarterly,* and *Written Communication.*

7. See Grabill and Pigg.

CHAPTER 1

1. The *Oxford English Dictionary* identifies the first use of "citizen science" as appearing in the journal *Technological Review* in 1989. See "Citizen Science."

2. See Fleming 23–54 on the storm controversy.

3. The navigational law codiscovered by Coffin and Ferrell is known as Buys Ballot's law. See "James Henry Coffin."

4. The National Weather Service's Cooperative Observer Program (COOP) still relies on volunteers to record measurements.

5. Unlike today, there was not a single Audubon Society. Instead, there was a confederation of Audubon Societies. See Stinson.

6. Stewart identifies the first appearance in E. H. Perkins's article "Some Results of Bird Lore's Christmas Bird Census" (1914). See Perkins.

7. The databases included *Ecology Abstracts, Environmental Science and Pollution Management, Pollution Abstracts, Water Resources, Web of Science,* and *Wiley Online Library.* Given the early adoption of citizen science in environmental fields, I chose to focus on databases dedicated to these sciences. *Web of Science* and *Wiley Abstracts* include a broader spectrum of sciences.

8. A search of the databases in note 7 using the phrase "citizen science" reveals that articles discussing the subject appear infrequently (two or fewer articles a year) until 2002 and only begin to significantly rise in frequency after 2006. From 2006 to 2007, the number of articles in the *Environmental Science and Pollution Management* database tripled from 10 to 31. The database *Web of Science* shows a more gradual but nonetheless perceptible rise in the number of articles during the period between 2006 and 2008 from six to twelve articles.

9. A detailed assessment of what scientific and sociological researchers have said about these dimensions of citizen science is presented in chapter 4 in the section "Citizen Science and Its Outcomes: A Review of the Literature."

CHAPTER 2

1. "Lifeworld" is a term used by Habermas to describe knowledge that arises out of people's unconscious, organic, day-to-day engagement with the world and each other. See Habermas 119–52. Habermas opposes the lifeworld knowledge to "system" knowledge, which arises through the conscious and disciplined application of inductive evidence and deductive reason. See Habermas 153–97.

2. Lee Brasseur's work on Florence Nightingale's development of rose diagrams to show the risks of poor sanitation in military hospitals takes a step in this direction. Its goal, however, is not a self-conscious discussion of risk visualization. See Brasseur. Beverly Sauer in *The Rhetoric of Risk* (2003) also talks about the role of visual media such as gesture and FATALGRAMS in mine safety education and discourse. This discussion, however, is limited to private specialized risk communication rather than public representation of risk. See Sauer 166–75; 232–44.

3. In 1961 there was an accident at a government test reactor SL-1 in Idaho where three people were killed. *Time* magazine and the *New York Times* ran stories on it in January of the same year. See Finney and "Runaway Reactor." The second accident was a partial meltdown in the core of the Fermi nuclear reactor outside Detroit in October 1966, which was reported on a month after the accident in the *New York Times.* See Gamson and Modiglioni 14.

4. See *Escape from H-Bomb.*
5. See US Department of Defense *Fallout Protection* 13.
6. See Krugler 184.
7. Cloud maps appeared in the *New York Times* on April 30 (A11), May 2 (A8), and May 16 (A6). They appeared in the *Washington Post* on May 1 (A34) and May 3 (A1).
8. The *New York Times* ran a story "New York and New Jersey Report No Excess Radioactivity Despite Patterns of Winds" on 2 April 1979. See McFadden.
9. See US Department of Defense, *Fallout Protection* 5, 14.
10. See Suter 16.
11. See Entman 52; Fahmy 147–49; and Perlmutter.

12. Reactors 1–3 were online. Reactors 4–6 were offline for maintenance and re-fueling.

13. There were a total of ten uses of the bull's-eye overlay and cloud visuals. The ratio of bull's-eye overlays to cloud visuals was 7:3.

14. See Cox, Ericson, and Tse A11 and "Japan's Assessment" A12.

15. The *Post* covered, for example, the way radiation gets into the biosphere and the debate over evacuation zones mandated by the Japanese government and those proposed by the US Embassy.

16. See Berkowitz et al. as an example.

17. This estimate is based on calculations done by Hiayan Zhang, a London-based interactive designer and software engineer who was also building a radiation visualization website and sharing data with RDTN.org. She gives these estimates in a 24 March 2011 interview in the *Atlantic* online "When Crowdsourcing Meets Nuclear Power." See Zhang.

18. Pachube is currently Cosm.

19. Akiba is the owner of Freaklabs, which designs and manufactures custom-built wireless sensor devices. See his website freaklabs.org for more details.

20. Major updates to the Safecast map were announced on Safecast's blog (blog.safecast.org) on Aug. 10, 2011; Feb. 3, 2012; and June 16, 2012.

21. Safecast has changed their color coding for risk over time. Initially, they used a conventional green-to-red risk scale. However, their most recent visualization breaks with this conventional color scheme by using a blue-to-light-yellow scale.

22. Funabashi is the chairman of the Rebuild Japan Initiative Foundation and Kitzawa is its staff director. The foundation established the Independent Investigation Commission on the Fukushima Daiichi Nuclear Accident, which was tasked with thoroughly examining the causes of and response to the Fukushima nuclear accident.

23. See Shimbun, Friedman, and IAEA.

CHAPTER 3

1. The distinction between specialist and semispecialist literature here can be illustrated by considering the difference between original research articles published in *Nature* or *Science* and the accommodations of those articles in the same publication. While the audience of the former would be specialists in the subject matter, the audiences of the latter would include scientists as well as scientific journalists who do not directly participate in the area of specialization of the original article.

2. See Prelli for a comprehensive discussion of these virtues.

3. By semi-expert appeal, I mean an appeal to an arguer's status as an amateur, enthusiast, or hobbyist in an area of specialized knowledge.

4. This estimate is based on calculations done by Hiayan Zhang, a London-based interactive designer and software engineer who was also building a radiation visualization website and sharing data with RDTN.org. She gives these estimates in a 24 March 2011 interview in the *Atlantic* online "When Crowdsourcing Meets Nuclear Power." See Zhang.

5. In his book *Appeal to Popular Opinion*, Douglas Walton examines the historical uses of *ad populum* and identifies no fewer than five different subtypes of this ap-

peal. The two that are significant for this analysis are the "bandwagon argument," which asserts the truth of a proposition on the grounds that it is what most people believe. The second challenges the validity of mass opinion on the grounds that it is based on emotion or undisciplined reflection. See Walton 62–63.

6. See Gertz, Howard, O'Brien, and Prichep.

CHAPTER 4

1. Public Engagement with Science includes the fields of Science and Technology Studies, Sociology of Science, Science Communication, and Environmental Justice.

2. See Bäckstrand; Ottinger; and Lövbrand, Pielke Jr., and Beck.

3. These forms, known as B-44 forms, are housed in the National Climatic Data Center in Asheville, North Carolina.

4. See Pielke et al. "Documentation" and Pielke et al. "Unresolved."

5. At the time of this writing, this weblog was located at http://pielkeclimatesci.wordpress.com/.

6. Pielke moved from Colorado State to the University of Colorado, Boulder, in 2005. See "Roger A. Pielke."

7. The blog post is titled "A New Paper on the Differences between Recent Proxy Temperature and In-situ Near-Surface Air Temperatures." The comments quoted from this post were accessed using the Internet archiving site Wayback Machine (archive.org), because the comments current and prior are currently blocked on the website. See Pielke Sr. "A New Paper."

8. May 17 is the date on which Watts created the prototype website for surfacestations.org. This information appears on slide #13 "website" from his CIRES power point presentation of 29 August 2007. See Watts *A Hands On Study*.

9. See Watts "Another Milestone."

10. The last update of the Surface Stations website was 30 July 2012.

11. Though Pielke himself was not responsible for introducing the siting criteria, it is likely that some contact Watts had made through Pielke had informed him of the ranking system. The first evidence that Watts knew about and began to apply the ranking system appears in the blog post "Standards for Weather Stations Siting Using the New CRN" on *wattsupwiththat.com* 3 July 2007.

12. See Leroy.

13. The Climate Reference Network (CRN) is a system of pristinely sited sensors meant to serve as a reference against which data from less-well-sited sensors in the USHCN network could be checked.

14. These numbers are from a search of *Lexus Nexus* and *Proquest* newspapers between June 2007 and May 2009 using the keywords "Surface Stations," "Anthony Watts," and "Roger Pielke Sr."

15. Bill Steigerwald's article appeared on 17 June 2007 just a week and a half after the launch of surfacestations.org.

16. I am basing my claim about the political centricity of the newspaper on the voting patterns of Butte county California where Oroville is located. In the 2008 election, the county went for President Obama by a slim margin (50%–47%). In the 2012 election, it went for Romney by a similarly slim margin (50%–46%). I am assuming that the newspaper reflects the political affiliations of its readers, which are politically divided. See "2012 Presidential Election" and "2008 Presidential Election."

17. The pronoun "we" appears twice in the section but not in reference to Pielke (e.g., "This is a big difference especially when we consider that the concern over anthropogenic global warming was triggered by . . ." and "Yet here we had an official climate-monitoring station.")

18. Stases are resting places of argument and commonly include the categories of existence/fact, definition, cause, quality, and action.

19. Pielke stipulated this in his answers to an email interview I conducted with him on 3 Nov. 2012. My questions and his answer were posted on his blog *http:// pielkeclimatesci.wordpress.com/2012/11/07/interview-with-james-wynn-in-the-english -department-at-carnegie-mellon-university/*.

20. A note at the beginning of the paper states that it was submitted on 27 August 2009. See Menne, Williams, and Palecki.

21. NOAA and the NCDC already identified errors in temperature measurement statistics introduced by such phenomena as the switch from mercury to digital thermometers and the change in taking temperature readings from noon to morning. For a detailed discussion, see Menne, Williams, and Palecki 1.

22. These terms come from Perelman and Olbrechts-Tyteca's *New Rhetoric*. See Perelman and Olbrechts-Tyteca 83–93.

23. Kenneth Walker and Lynda Walsh explain that the *topos* of uncertainty can be used to undermine scientific ethos and promote or stymie political action. In this case, however, it opens up a space for research in science and in doing so gives priority to qualitative reasoning. See Walker and Walsh 9–10.

24. Approximately 87% of stations were checked and surveyed as opposed to 43% in Menne, Williams, and Palecki. See Revkin.

25. See Miller; Corburn; Ottinger and Cohen; and Brown, Morello-Frosch, and Zavestoski.

26. I am borrowing but repurposing this term from the sociologist Roger Pielke Jr., son of the climate scientist discussed in this chapter. He defines an "honest broker" as a scientist who offers technical alternatives to decision makers in a policy context without promoting one perspective or another. I am using it instead to describe the role a rhetorical analyst might play in pointing out alternatives of discourse and argument and the possible repercussions they might have for the relationships between laypersons and citizens without suggesting what decisions they should make in their public or technical address of one another. For an explanation of the "honest broker" see Roger Pielke Jr. 1–3.

CHAPTER 5

1. I am using the term *public* here to refer to someone who is not a member of the decision-making elite.

2. Barking and Dagenham, Greenwich, Hackney, Newham, Tower Hamlets, and Waltham Forest.

3. The Thames Gateway project included development in a number of locations in south and east London including the boroughs of Newham (Royal Docks and Stratford), Lewisham (Deptford), Barking and Dengham (Barking), and Greenwich (Greenwich Peninsula). See "Thames Gateway."

4. This project was funded from 2007–2009. See "Mapping Change" *London 21*.

5. These included Pepys Estate in the borough of Lewisham, the Royal Docks

area of the borough of Newham, the Archway area of the borough of Islington, the Marks Gate neighborhood of Barking and Dengham borough, and the Hackney Wick community in the borough of Hackney. See "Pilot Groups."

6. The readings with the Mapping Change for Sustainable Communities project took place in January and February of 2008.

7. See appendix A.

8. Fahnestock and Secor label these the stases of definition, cause, evaluation, and proposal.

9. *Rhetoric Society Quarterly, Rhetoric and Public Affairs, Environmental Communication, Technical Communication Quarterly,* and *Written Communication*

10. There is not at the time of writing any rhetorical scholarship dealing with citizen science.

11. These parties include residents, their environmental justice advocates, and federal and state environmental agencies.

12. According to Ottinger, she interviewed five regulators from either the EPA or the Louisiana Department of Environmental Quality and ten high-ranking engineers and scientists at Shell Chemical. Finally, she was a participant observer in two environmental justice groups: Communities for a Better Environment and Louisiana Bucket Brigade. See Ottinger 247.

13. See Haklay, Francis, Whitaker 26 and Pepys Community 8.

14. The participants included Luciana Dualibe, Caroline Fox, James Davies, and Dalva James. See Pepys Community 8.

15. These measurements are the averages of three individual readings taken one minute apart from one another. The total number of individual readings is 1155.

16. The Environmental Agency is the United Kingdom's equivalent of the EPA.

17. Caroline Fox did not present.

18. See appendix B.

19. dBA in this case stands for the equivalent continuous A-weighted sound pressure level—that is the level audible to the human ear—rated in decibels over some designated period of time. It is important to note that in all instances we are referring to averages of sound levels over time not individual sound events. See "Noise Mapping" 1 and "Frequency Weightings."

20. dB LAeq is equivalent to dBA. See "Frequency Weightings."

21. I am using a letter instead of the name of participant here to protect their anonymity.

22. See appendix C.

23. The categories were airplane, birds, lorry [truck], other, scrapyard, traffic, and train.

24. The scrapyard received approximately 108 and the traffic 110. The next closest was airplane with approximately 38 identifications. See appendix D.

25. I use "at most" here because the speaker makes no clear locational reference so it is not completely certain that she is even referring to this area. It could be the case that she ignores these readings completely, which strengthens my conclusion that the high density areas where we would expect sound from the scrapyard is of little importance to her.

26. These include Lund, Jepsen, and Simonsen; World Health Organization, "Occupational and Community Noise"; and "Loud Noises."

27. The measurements were made on the 26 and 29 of January 2008, the first at 1:17 P.M. and the second at 12:24 P.M.

28. There are three sources that point to Alexander as the person who initiated contact with the Mapping Change for Sustainable Communities project. One of them is an interview I conducted with Muki Haklay. See Haklay. Another is an audio file of the 15 May 2008 meeting sent to me by Haklay in which Alexander herself says, "One of the reasons this work [the sound mapping] is actually taking place here is because I said I want it to happen." The third source is the *Pepys Community Forum Annual Report*. See Pepys 8.

29. These included Millennium Quay, Convoys Wharf, Paynes/Borthwick Wharf, Deptford Wharves, Creekside Charrette, Cannon Wharf/Plough Way, and Marine Wharf.

30. These included Convoys Wharf, Deptford Wharves, Cannon Wharf/Plough Way, and Marine Wharf. See Ipsos MORI 8–10.

31. This number was arrived using a keyword search of the PDF for uses of the words *noise, scrap,* and *scrapyard.*

32. *The Wharves, Deptford Statement of Community Involvement* documents three consultations with Herlitz on 3 Dec. 2008, 13 Aug. 2009, and 15 Sep. 2009. See Soundings 219–25.

33. 13 Aug. 2009. See Soundings 222–25.

34. This meeting took place on 16 July 2009. See Soundings 242.

35. In this case, the firm is Soundings, a London-based stakeholder consultancy firm.

36. See Cadman, *Pepys.*

37. According to Sieber (2006), PPGIS "pertains to the use of geographic information systems (GIS) to broaden public involvement policymaking was well as to the value of GIS to promote the goals of nongovernmental organizations, grassroots groups, and community based organizations" (491).

Works Cited

Aaron, Jacob. "Japan's Crowdsourced Radiation Maps." *Newscientist.com*. New Scientist, 30 Mar. 2011. Web. 9 Dec. 2011.

"About eBird." *Ebird.org*. Cornell Lab of Ornithology and National Audubon Society, n.d. Web. 6 Aug. 2013.

"About SETI@home." *Seti@home.berekely.edu*. SETI@home, n.d. Web. 12 Mar. 2013.

"About the Christmas Bird Count." *Audubon.org*. Audubon Society, n.d. Web. 12 Dec. 2012.

Aerial Monitoring Results Fukushima Daiichi, Japan. Map. *Energy.gov*. U.S. Department of Energy, 25 Mar. 2011. Web. 1 Aug. 2012. PowerPoint.

Akiba, [Christopher Wang]. "Hacking a Geiger Counter in Nuclear Japan." *Freaklabs .org*. Freaklabs, 24 Mar. 2011. Web. 4 Jan. 2012.

———. "Re: [Safecast Jpn] Experts: Leave Radiation Checks to Us." *Groups.google.com*. Safecast-Japan, 28 May 2011. Web. 27 July 2012.

———. Telephone interview. 25 July 2012.

———. "Thanks for all the support! And What's Coming up." *Freaklabs.org*. Freaklabs, 16 Mar. 2011. Web. 8 Mar. 2012.

Alvarez, Marcelino. "72 Hours from Concept to Launch: RDTN.org." *Uncorkedstudios .com*. Uncorked Studios, 21 Mar. 2011. Web. 15 Dec. 2011.

———. Telephone interview. 19 July 2012.

Amerisov, Alexander. "A Chronology of Soviet Media Coverage." *Bulletin of the Atomic Scientist* 43.1 (1986): 54–56. Print.

Anderson, Katherine. *Predicting the Weather: Victorians and the Science of Meteorology*. Chicago: Chicago UP, 2005. Print.

"Anthony Watts (blogger)." *Wikipedia.org*. Wikipedia, 28 Sept. 2012. Web. 30 Sept. 2012.

Aoki, Kazumasa. "Wanted: The Right to Relocate." *Fukushima.greenaction-japan.org*. Green Action Japan, 3 Aug. 2011. Web. 10 July 2013.

ApSimon, Helen, and Julian Wilson. "Tracking the Cloud from Chernobyl." *New Scientist* 111 (1986): 42–45. Print.

Arbib, Robert. "'Ideal Model' Christmas Bird Counts: A Start in 1982–83." *American Birds* 36.2 (1982): 146–148. Web. 12 Dec. 2012.

Aristotle. *Rhetoric*. Trans. W. Rhys Roberts. New York: Random House, 1984. Print.

Associated Press. "Radiation Effects." Map. 4 Mar. 1955. *AP Images.com*. 9 May 2011. Web. 15 Jul. 2011.

Atomic Energy Commission. *WASH-740: Theoretical Possibilities and Consequences of Major Nuclear Accidents in Large Nuclear Power Plants.* Washington: GPO, 1957.

Bäckstrand, Karin. "Civic Science for Sustainability: Reframing the Roles of Experts, Policy-Makers, and Citizens in Environmental Governance." *Global Environmental Politics* 3.4 (2003): 24–41. Web. 16 Mar. 2012.

Beck, Ulrich. *Risk Society: Towards a New Modernity.* Thousand Oaks: Sage, 1992. Print.

Berkowitz, Bonnie, et al. *Path of the Plume.* Map. *Washington Post* 16 Mar. 2011: A10. Microfilm.

Birdsell, David, and Leo Groarke. "Outlines of a Theory of Visual Argument." *Argumentation and Advocacy* 43 (2007): 103–113. Print.

Blair, William. "U.S. H-Bomb Test Put Lethal Zone at 7,000 sq. Miles." *New York Times* 16 Feb. 1955: 1+. *ProQuest Historical Newspapers.* Web. 11 June 2012.

Bloch, Matthew, et al. *Map of the Damage from the Japanese Earthquake.* Map. *Nytimes.com.* New York Times, 10 Apr. 2011. Web. 12 June 2012.

Bogost, Ian. *Persuasive Games: The Expressive Power of Videogames.* Cambridge: MIT P, 2007. Print.

Bohlen, Celestine. "Soviet Nuclear Accident Sends Radioactive Cloud Over Europe." *Washington Post* 29 Apr. 1986: A1+. Microfilm.

Bolter, David, and Richard Grusin. *Remediation: Understanding New Media.* Cambridge: MIT P, 1999. Print.

Bonner, Sean. "Alpha, Beta, Gamma." *Safecast.org.* Safecast, 5 May 2011.Web. 6 July 2012.

———. "First RDTN Sensor Deployed." *Safecast.org.* Safecast, 14 Apr. 2011. Web. 3 July 2012.

———. "First Safecast Mobile Recon." *Safecast.org.* Safecast, 24 Apr. 2011. Web. 3 July 2012.

———. "RDTN.org: Crowdsourcing and Mapping Radiation Levels." *Boingboing.net.* Boingboing, 19 Mar. 2011. Web. 17 Dec. 2011.

———. Telephone interview. 31 July 2012.

Bonney, Rick, et al. "Citizen Science: A Developing Tool for Expanding Science Knowledge and Scientific Literacy." *Bioscience* 59.11 (2009): 977–984. Web. 26 Sept. 2012.

———. *Public Participation in Scientific Research Defining the Field and Assessing Its Potential for Informal Science Education.* A CAISE Inquiry Group Report. Washington DC: Center for Advancement of Informal Science Education (CAISE), 2009. Web. 29 Oct. 2012.

Brasseur, Lee. "Florence Nightingale's Visual Rhetoric in the Rose Diagrams." *Technical Communication Quarterly* 14.2 (2005): 161–82. Print.

Brooksher, Dave. "RDTN.org Peer Reviews Crowdsource Radiation Data." *Farwest.fm.* FarWest.FM, 1 Apr. 2011. Web. 5 Jan. 2011.

Brown, Phil, Rachel Morello-Frosch, and Stephen Zavestoski. *Contested Illnesses: Citizens, Science, and Health Social Movements.* Berkeley: U of California P, 2011. Print.

Browne, Malcolm. "Winds Blow Fallout to Southern Europe." *New York Times* 2 May 1986: A8. *ProQuest Historical Newspapers.* Web. 15 Jun. 2012.

Bull's-eye Overlay of the Three Mile Island Accident. Map. *New York Times* 31 Mar. 1979: A1.

Cadman, Malcolm. "Objection to Planning Application, The Wharves, Oxestalls Road, Deptford." Message to the London Borough of Lewisham Planning Commission. 28 June 2010. E-mail.

——. *Pepys Community Forum Archive.* "Environmental Justice Project." *Mcad.demon .co.uk.* Malcolm Cadman, n.d. Web. 3 May 2013.

Chapman, Frank. "The AOU and the Audubon Societies." *Bird Lore* 2.5 (1900): 161–61. Print.

——. "A Christmas Bird Census." *Bird Lore* 2.6 (1900): 192. Print.

"Chernobyl: Half Hidden Disaster." Editorial. *Washington Post* 1 May 1986: A22. Microfilm.

"Chernobyl's Other Cloud." Editorial. *New York Times* 30 Apr. 1986: A30. *ProQuest Historical Newspapers.* Web. 15 Jun 2012.

Cicero. *De Inventione.* Trans. H.M. Hubbell. Cambridge: Harvard UP, 2000. Print.

"Citizen Science." Def. C3. OED Online. Oxford UP, Mar. 2015. Web 30 Apr. 2015.

"'Citizen Science' Takes Off as Residents Tackle Neighborhood Noise." *London21.org.* London 21, n.d. Web. 16 Mar. 2012.

Coffin, James Henry. *Winds of the Globe or the Laws of Atmospheric Circulation over the Surface of the Earth.* Washington DC: Smithsonian, 1857. *Google Book Search.* Web 12 Dec. 2012.

——. *Winds of the Northern Hemisphere.* Washington DC: Smithsonian, 1852. *Google Book Search.* Web 12 Dec. 2012.

Cohn, Jeffrey. "Citizen Science: Can Volunteers Do Real Research?" *Bioscience* 58.3 (2008): 192–97. Web. 26 Sept. 2012.

Collins, Henry, and Robert Evans. *Rethinking Expertise.* Chicago: U of Chicago P, 2007. Print.

Cook, David. "Area Surrounding Three Mile Island Nuclear Plant." Map. *Washington Post* 31 Mar. 1979: A8. Microfilm.

Cook, Gareth. "How Crowdsourcing Is Changing Science." *Bostonglobe.com.* Boston Globe, 11 Nov. 2011. Web. 10 Dec. 2011.

Cooper, Caren, et al. "Citizen Science as a Tool for Conservation in Residential Ecosystems." *Ecology and Society* 12.2 (2007): 1–11. Web. 26 Sept. 2012. PDF file.

Corburn, Jason. *Street Science: Community Knowledge and Environmental Health Justice.* Cambridge: MIT P, 2005. Print.

Cox, Amanda, Matthew Ericson, and Archie Tse. "The Evacuation Zones around the Fukushima Plant." Map. *New York Times* 18 Mar. 2011: A11. *ProQuest Historical Newspapers.* Web. 12 Jun. 2011.

——. "The Evacuation Zones around the Fukushima Plant." Map. *Nytimes.com.* New York Times, 25 Mar. 2011. Web. 6 Jun. 2012.

Daston, Lorraine. *Classical Probability in the Enlightenment.* Princeton: Princeton UP, 1988. Print.

Davey, Christopher, and Roger Pielke Sr. "Microclimate Exposures of Surface-based Weather Stations: Implications for the Assessment of Long Term Temperature Trends." *Bulletin of the American Meteorological Society* 86.4 (2005): 497–504. Web. 9 Oct. 2012.

Davisson, Amber. "Beyond the Borders of Red and Blue States: Google Maps as a Site of Rhetorical Invention in the 2008 Presidential Election." *Rhetoric and Public Affairs* 14.1 (2011): 101–24. Print.

Dickinson, Janis, and Rick Bonney. Introduction. *Citizen Science: Public Participation in Environmental Research.* Eds. Janis Dickinson and Rick Bonney. Ithaca: Cornell UP, 2012. 1–14. Print.

Dickinson, Janis, Benjamin Zuckerberg, and David Bonter. "Citizen Science as an Ecological Research Tool: Challenges and Benefits." *Annual Review of Ecology, Evolution, and Systematics* 41 (2010): 149–72. Web. 26 Sept. 2012.

Dobbs, Michael. "Soviet Drive for New Image Put in Jeopardy by Accident." *Washington Post* 1 May 1986: A1. Microfilm.

Dorman, William, and Daniel Hirsch. "The U.S. Media's Slant." *Bulletin of the Atomic Scientist* 43.1 (1986): 54–56. Print.

Drew, Michael. "Path of Fallout." Map. *Washington Post* 1 May 1986: A34. Microfilm.

Dunn, Erica H., et al. "Enhancing the Value of the Christmas Bird Count." *The Auk* 122.1 (2005): 338–46. Web. 12 Dec. 2012.

Endres, Danielle. "Science and Public Participation: An Analysis of Public Scientific Argument in the Yucca Mountain Controversy." *Environmental Communication* 3.1 (2009): 49–75. Print.

Entman, Robert. "Framing toward a Clarification of a Fractured Paradigm." *Journal of Communication* 43.1 (1993): 51–58. Web. 9 Jul. 2012.

Epstein, Steven. "The Construction of Lay Expertise: AIDS Activism and the Forging of Credibility in the Reform of Clinical Trials." *Science, Technology, and Human Values* 20.4 (1995): 408–37. Web. 9 Apr. 2012.

Escape from H-Bomb: St. Louis County and City. Map. St. Louis: Office of Civil Defense, 1955. N.d. Web. 11 June 2012.

Ewald, David. "Open Dialogue." *Safecast.org*. Safecast, 23 Mar. 2011. Web. 5 Jul. 2012.

"Experts: Leave Radiation Checks to Us." *Groups.google.com*. Safecast-Japan, 28 May 2011. Web. 27 July 2012.

Fabj, Valeria, and Matthew Sobnosky. "AIDS Activism and the Rejuvenation of the Public Sphere." *Argument and Advocacy* 32.4 (1995): 163–84. Print.

Fahmy, Shahira. "They Took It Down: Exploring Determinants of Visual Reporting in the Toppling of the Saddam Statue in National and International Newspapers." *Mass Communication and Society* 10.2 (2007): 143–70. Web. 21 Jun. 2012.

Fahnestock, Jeanne. "Accommodating Science: The Rhetorical Life of Scientific Facts." *Written Communication* 15.3 (1998): 330–50. Print.

Fahnestock, Jeanne, and Marie Secor. *A Rhetoric of Argument*. New York: McGraw Hill, 2004. Print.

Fall, Souleymane, et al. "Analysis of the Impacts of Station Exposure on the U.S. Historical Climatology Network Temperatures and Temperature Trends." *Journal of Geophysical Research* 116 (2011): 1–15. Web. 26 Sep. 2012.

Farrell, Thomas, and Thomas Goodnight. "Accidental Rhetoric: The Root Metaphors of Three Mile Island." *Landmark Essays on Rhetoric and the Environment*. Ed. Craig Waddell. Mahwah, New Jersey: Lawrence Erlbaum, 1998. 75–105. Print.

Finney, John. "Atom Aides Scan Effect of Blast." *New York Times* 5 Jan 1961: 19. *ProQuest Historical Newspapers*. Web. 11 June 2012.

Fischer, Frank. *Citizens, Experts, and the Environment: The Politics of Local Knowledge*. Durham: Duke UP, 2000. Print.

Fisher, Walter. *Human Communication as Narration: Toward a Philosophy of Reason, Value, and Action*. Columbia: U of South Carolina P, 1987. Print.

Fitzpatrick, John. Afterward. *Citizen Science: Public Participation in Environmental Research*. Eds. Janis Dickinson and Rick Bonney. Ithica: Cornell UP, 2012. 235–40. Print.

Fleming, James Roger. *Meteorology in America*. Baltimore: Johns Hopkins UP, 1990. Print.

"FoldIt." *Wikipedia.org*. Wikipedia. 5 Aug. 2010. Web. 12 Mar. 2013.

Foreman, Edward. "Report of the General Assistant with Reference to the Meteorological Correspondence." *Sixth Annual Report of the Board of Regents of the Smithsonian Institution*. Washington, DC: Robert Armstrong, 1852. *Google Book Search*. Web 15 Dec. 2012. PDF file.

"Frequency Weightings." *Acousticglossary.co.uk*. Gracey & Associates, n.d. Web. 10 June 2013.

Friedman, Sharon. "Three Mile Island, Chernobyl, and Fukushima: An Analysis of Traditional and New Media Coverage of Nuclear Accidents and Radiation." *Bulletin of the Atomic Scientists* 67.5 (2011): 55–65. Web. 21 Jan. 2012.

Funabashi, Yoichi, and Kay Kitazawa. "Fukushima in Review: A Complex Disaster, a Disastrous Response." *Bulletin of the Atomic Scientists* 0.0 (2012): 1–13. Web. 5 Aug. 2012.

Furno, Dick. *Radiation Was Detected as Far as Harrisburg*. Map. *Washington Post* 29 Mar. 1979: A7. Microfilm.

Gamson, William, and Andre Modigliani. "Media Discourse and Public Opinion on Nuclear Power: A Constructionist Approach." *The American Journal of Sociology* 95.1 (1989): 1–37. Print.

Gertz, Emily. "Got iGeigie? Radiation Monitoring Meets Grassroots Mapping." *Onearth .org*. Onearth, 20 Apr. 2011. Web. 5 Jan. 2012.

Gibbons, Michelle. "Seeing the Mind in the Matter: Functional Brain Imaging as Framed Visual Argument." *Argument and Advocacy* 43 (2007): 175–88. Web. 12 Mar. 2014.

Grabill, Jeffrey, and Michelle Simmons. "Toward a Critical Rhetoric of Risk Communication: Producing Citizens and the Role of Technical Communicators." *Technical Communication Quarterly* 7.4 (1998): 415–41. Print.

Grabill, Jeffrey, and Stacey Pigg. "Messy Rhetoric: Identity Performance as Rhetorical Agency in Online Public Forums." *Rhetoric Society Quarterly* 42.2 (2012): 99–119. Web. 15 Aug. 2012.

Gray, Jim. "Distributed Computing Economics." *Research.microsoft.com*. Microsoft, Mar. 2003. Web. 14 Mar. 2014.

Great Britain. Department of Culture, Media, and Sport. *Creating a Lasting Legacy from the 2012 Olympic and Paralympic Games*. *Gov.uk*. 20 Feb. 2013. Web. 4 June 2013.

Greater Boston Civil Defense Manual. Boston: Civil Defense Authority, 1952. *Internet Archive*. 11 June 2012. PDF file.

Gross, Alan. "Toward a Theory of Verbal-Visual Interaction: The Example of Lavoisier." *Rhetoric Society Quarterly* 39.2 (2009): 147–69. Print.

Gross, Alan, Joseph Harmon, and Michael Reidy. *Communicating Science: The Scientific Article from the 17th Century to the Present*. Oxford: Oxford UP, 2002. Print.

Guyot, Arnold Henry. *Directions for Meteorological Observations and the Registry of Periodical Phenomena*. Washington, DC: The Smithsonian Institution, 1858. *Google Book Search*. Web 12 Dec. 2012. PDF file.

Gwertzman, Bernard. "Plume of Radioactive Material Spread from Accident at Pripyat." *New York Times* 29 Apr. 1986: A1. Map. *ProQuest Historical Newspapers*. Web. 22 May 2012.

Habermas, Jürgen. *The Theory of Communicative Action*. Vol. 2. Boston: Beacon P, 1987. Print.

Hagan, Susan. "Visual/Verbal Collaboration in Print: Complementary Differences,

Necessary Ties and an Untapped Rhetorical Opportunity." *Written Communication* 24.1 (2007): 49–83. Print.

Haklay, Muki. Skype interview. 28 May 2013.

Haklay, Muki, Louise Francis, and Colleen Whitaker. "Mapping Noise Pollution." *GIS-Professional* 23 (2008): 26–28. Print.

Hansen, James, et al. "A Closer Look at United States and Global Surface Temperature Change." *Journal of Geophysical Research* 106 (2001): 1–13. Web. 11 Nov. 2012.

Hartelius, Johanna E. *The Rhetoric of Expertise*. Lanham, MD: Lexington Books, 2011. Print.

Heartland Institute. "Global Warming: Not a Crisis." *Heartland.org*. Heartland Institute, n.d. Web. 30 Sep. 2012.

Herndl, Carl, and Stuart Brown. "Introduction: Rhetorical Criticism and the Environment." *Green Culture*. Madison: U of Wisconsin P, 1996. Print.

Herschel, John. *A Preliminary Discourse on the Study of Natural Philosophy*. 1830. New York: Johnson Reprint, 1966. Print.

Hickey, Joseph. "Letter to the Editor." *The Wilson Bulletin* 67.2 (1955): 144–45. Web. 12 Dec. 2012.

"History." *Safecast.org*. Safecast, n.d. Web. 9 Dec. 2011.

Howard, Alex. "Citizen Science, Civic Media and Radiation Data Hints at What's to Come." *Radar.oreilly.com*. O'Reilly Radar, 29 June 2011. Web. 3 Jan. 2011.

International Atomic Energy Agency. *IAEA International Fact Finding Mission of the Nuclear Accident Following the Great East Japan Earthquake and Tsunami*. Prime Minister of Japan and His Cabinet, June 1 2011. Web. 8 June 2011.

Ipsos MORI and Urban Practitioners. *North Deptford Consultation. Lewisham.gov.uk*. London Borough of Lewisham, Feb. 2009. Web. 30 May 2013. PDF file.

Irwin, Alan. *Citizen Science: A Study of People, Expertise and Sustainable Development*. New York: Routledge, 1995. Print.

Ishikawa, Yuki. "Calls for Deliberative Democracy in Japan." *Rhetoric and Public Affairs* 5.2 (2002): 331–45. Print.

Ito, Joi. "A Rock in One Hand Cell Phone in the Other." MIT-Knight Civic Media Conference. Cambridge, Mass. 24 Jun. 2011. Web. 30 Jul. 2012. Presentation.

ITWorks. N.d. Web. 11 Nov. 2012.

Jamail, Dahr. "Citizen Group Tracks Down Japan's Radiation." *Aljazeera.com*. Aljazeera, 10 Aug. 2011. Web. 3 Jan. 2012.

James. "Crowdsourcing Radiation Monitoring." *Mapt.com*. Mapt, 21 Mar. 2011. Web. 30 June 2011.

"James Henry Coffin." *Wikipedia.org*. Wikipedia, 20 Apr. 2012. Web. 12 Dec. 2012.

"Japan's Assessment of Radiation around Plant." Map. *New York Times* 19 Mar. 2011: A12. Microfilm.

Katz, Stephen, and Carolyn Miller. "The Low-Level Radioactive Waste Siting Controversy in North Carolina: Toward a Rhetorical Model of Risk Communication." *Green Culture*. Eds. Carl Herndl and Stuart Brown. Madison: U of Wisconsin P, 1996. 111–40. Print.

Keränen, Lisa. "Concocting Viral Apocalypse: Catastrophic Risk and the Production of Bio(in)security." *Western Journal of Communication* 75.5 (2011): 451–72. Web. 11 May 2012.

———. *Scientific Characters: Rhetoric, Politics, and Trust in Breast Cancer Research*. Tuscaloosa: U of Alabama P, 2010. Print.

Khatib, Firas, et al. "Crystal Structure of a Monomeric Retroviral Protease Solved by Protein Folding Game Players." *Nature.com*. Nature Structural and Molecular Biology, 18 Sept. 2011. Web. 19 Sept. 2011.

Kinsella, William. "Public Expertise: A Foundation for Citizen Participation in Energy and Environmental Decisions." *Communication and Public Participation in Environmental Decision Making*. Eds. Stephen Depoe, John Delicath, and Marie-France Aepli Elsenbeer. New York: SUNY P, 2004.

Kitzinger, Jenny. *The Media and Public Risk*. Great Britain. Risk and Regulation Advisory Council. Oct. 2009. Web. 7 Sep. 2011.

Knox, Richard. "'Citizen Scientists' Crowdsource Radiation Measurements in Japan." *Npr.org*. National Public Radio, 24 Mar. 2011. Web. 8 Dec. 2011.

Kostelnick, Charles, and Michael Hassett. *Shaping Information: The Rhetoric of Visual Conventions*. Carbondale: U of Illinois P, 2003. Print.

Kress, Gunther, and Theo Van Leeuwen. *Reading Images: The Grammar of Visual Design*. 2nd ed. London: Routledge, 2006. 16–44. Print.

Krieger, Daniel. "Monitoring the Monitors." *Slate.com*. Slate Magazine, 16 June 2011. Web. 21 Jan. 2012.

Krugler, David. *This Is Only a Test: How Washington D.C. Prepared for Nuclear War*. New York: Palgrave Macmillan, 2006. Print.

LaFrance, Adrienne. "After Tracking Radiation Levels in Fukushima, Safecast Is Measuring Air Quality in the States." *Niemanlab.org*. Nieman Journalism Lab, 21 Sept. 2012. Web. 22 May 2014.

Lawton, Keith, and Stephanie Briscoe, comp. *Novel Approaches to Waste Crime: A Report of Three Waste Crime Prevention Pilots*. European Pathway to Zero Waste. Reading, UK: Environment Agency, 2012.Web. 17 June 2013. PDF file.

Leach, Melissa, and Ian Scoones. "Science and Citizenship in a Global Context." *Science and Citizens: Globalization and the Challenge of Engagement*. Ed. Melissa Leach, Ian Scoones, and Brian Wynne. London: Zed Books, 2005. Print.

Lee, Gary. "Soviets Say Clean up Underway." *Washington Post* 1 May 1986: A1. Microfilm.

Leitsinger, Miranda. "Japanese Government Responds to Citizen Scientists' Radiation Mapping." *Worldblog.msnbc.com*. NBC News, 14 Jul. 2011. Web. 8 Aug. 2011.

———. "Japan's Citizen Scientists Map Radiation, DIY-style." *Worldblog.msnbc.com*. NBC News, 12 Jul. 2011. Web. 8 Aug. 2011.

Leroy, Michel. "Classification d'un Site." *Note Technique* 35. Trappes: Director des Systemes d'Observation Meteo-France, 1999. Web. 22 July 2013.

Lewis, Flora. "Moscow's Nuclear Cynicism." Editorial. *New York Times* 1 May 1986: A27. *ProQuest Historical Newspapers*. Web. 14 Jun. 2012.

"Loud Noises 'Bad for Heart.'" *BBC.co.uk*. British Broadcasting Corporation, 24 Nov. 2005. Web. 14 June 2013.

Lövbrand, Eva, Roger Pielke Jr., and Silke Beck. "A Democracy Paradox in Studies of Science and Technology." *Science, Technology, and Human Values* 36.4 (2011): 474–96. Web. 4 Oct. 2012.

Luke, Timothy. "Chernobyl: The Packaging of Transnational Ecological Disaster." *Critical Studies in Mass Communication* 4 (1987): 351–75. Print.

Lund, Soren, Gitte Jepsen, and Leif Simonsen. "Effect of Long-term, Low-level Noise Exposure on Hearing Thresholds, DPOAE and Suppression of DPOAE in Rats." *Noise and Health* 3.12 (2001): 33–42. Web.16 June 2013.

Lynch, Patrick, and Sarah Horton. *Web Style Guide.* 3rd ed. New Haven: Yale UP, 2008. Print.

Lyne, John, and Henry F. Howe. "The Rhetoric of Expertise: E.O. Wilson and Sociobiology." *Quarterly Journal of Speech* 76 (1990): 134–51. Web. 27 May 2014.

Lyons, Richard. "Children Evacuated." *New York Times* 31 Mar. 1979: A1. *ProQuest Historical Newspapers.* Web. 18 July 2011.

"Making Maps Work for Communities." *UCL Enterprise.* University College London, n.d. Web 6 May 2013.

Map of Radiation Measurements by Greenpeace Team. Googlemaps.com. Google, 27 Mar. 2011. Web. 2 July 2013.

"Mapping Change for Sustainable Communities." *London 21.org.* London 21, n.d. Web. 5 June 2013.

"Mapping Change for Sustainable Communities." *UrbanBuzz.* UrbanBuzz, n.d. Web. 2 Mar. 2013.

Mathews, Tom, et al. "Nuclear Accident." *Newsweek.* 9 Apr. 1979: 24–33. Print.

Matsutani, Minoru. "Experts: Leave Radiation Checks to Us Laypersons Just Spread Fear with Inaccurate Readings, They Say." *Japantimes.co.jp.* Japan Times, 28 May 2011. Web. 17 Dec. 2011.

McCallie, Ellen, et al. *Many Experts, Many Audiences: Public Engagement with Science and Informal Science Education.* Washington, DC: Center for Advancement of Informal Science Education, 2009. Web. 29 Oct. 2012.

McFadden, Robert. "New York and New Jersey Report No Excess Radioactivity Despite Patterns of Winds." *New York Times* 2 Apr. 1979: A16. *ProQuest Historical Newspapers.* Web. 14 Jun. 2012.

"Meltdown at Chernobyl." Editorial. *Washington Post.* 30 Apr. 1986: A24. Microfilm.

Menne, Matthew, Claude Williams, and Michael Palecki. "On the Reliability of the U.S. Surface Temperature Record." *Journal of Geophysical Research* 115 (2010): 1–9. Web. 28 Sept. 2012.

Menne, Matthew, Claude Williams, and Russell Vose. "The U.S. Historical Climatology Network Monthly Temperature Data, Version 2." *Bulletin of the American Meteorological Society* 90.7 (2009): 993–1007. Web. 28 Sept. 2012.

Miller, Carolyn R. "The Presumption of Expertise: The Role of Ethos in Risk Analysis." *Configurations* 11.2 (2003): 163–202. Print.

National Academy of the Sciences. *Adequacy of Climate Observing Systems.* Washington, DC: National Academy P, 1999. Web. 11 Nov. 2012.

National Oceanic and Atmospheric Administration. *Climate Reference Network Site Information Handbook.* Asheville: US Department of Commerce, National Climatic Data Center, 2002. Web. 10 Oct. 2012.

———. *United States Climate Reference Network Part of NOAA's Environmental Real-time Observations Network FY 2005.* Asheville: US Department of Commerce, National Climatic Data Center, 2005. Web. 22 July 2013.

"Noise Mapping Toolkit." *Mappingforchange.org.uk.* Mapping for Change, 5 May 2009. Web. 3 May 2013. PDF file.

O'Brien, Miles. "Safecast Draws on Power of the Crowd to Map Japan's Radiation." *Pbs.org.* PBS News Hour, 10 Nov. 2011. Web. 8 Dec. 2011.

Oldenburg, Henry. "Introduction." *Philosophical Transactions* 1 (1665): 1. Web. 31 July 2013.

Olson, Ryan. "Watts' up? Spotlight Shines on Local Weatherman's Latest Research." *Orovillemr.com*. Oroville Mercury-Register, 29 June 2007. Web. 20 Nov. 2012.

"One Million Volunteers." *Zooniverse.org*. Zooniverse, 14 Feb. 2014. Web. 22 May 2014.

Onishi, Norimitsu, and Martin Fackler. "Japan Held Nuclear Data, Leaving Evacuees in Peril." *Heraldtribune.com*. Sarasota Herald-Tribune, 8 Aug. 2011. Web. 9 Dec. 2011.

Ottinger, Gwen. "Buckets of Resistance: Standards and the Effectiveness of Citizen Science." *Science, Technology, and Human Values* 35.2 (2010): 244–70. Web. 27 Sept. 2012.

Ottinger, Gwen, and Benjamin Cohen, eds. *Technoscience and Environmental Justice: Expert Cultures in Grassroots Movement*. Cambridge: MIT P, 2011. Print.

"Path of Airborne Radiation." Map. *New York Times* 2 May 1986: A8.

Pepys Community Forum. *Pepys Community Forum Annual Report*. *Charitycommission.uk.gov*. Charity Commission, 31 Mar. 2009. Web. 2 Mar. 2013.

Perelman, Chiam. *The Realm of Rhetoric*. Notre Dame: U of Notre Dame P, 1982. Print.

Perelman, Chiam, and Lucie Olbrechts-Tyteca. *The New Rhetoric*. Notre Dame: U of Notre Dame P, 1971. Print.

Perkins, E. H. "Some Results of Bird Lore's Christmas Bird Census." *Bird Lore* 16 (1914): 13- 18. Print.

Perlmutter, David. *Visions of War*. New York: St. Martin's P, 1999. Print.

"Phenology." *Wikipedia.org*. Wikipedia. 20 Nov. 2006. Web. 5 Jan. 2013.

Pielke, Roger, Jr. *The Honest Broker: Making Sense of Science in Policy and Politics*. Cambridge: Cambridge UP, 2007. Print.

Pielke, Roger, Sr. Interview. 3 Nov. 2012. E-mail.

———. "A New Paper on the Differences between Recent Proxy Temperature and In-situ Near- surface Air Temperatures." *Climate Science: Roger Pielke Sr.* 4 May 2007. Web. 13 Nov. 2012.

———. "Re: A New Paper on the Differences between Recent Proxy Temperature and In-situ Near- surface Air Temperatures." *Climate Science: Roger Pielke Sr.* 4 May 2007. Web. 13 Nov. 2012.

Pielke, Roger, Sr., et al. "Documentation of Uncertainties and Biases Associated with Surface Temperature Measurement Sites for Climate Change Assessment." *Bulletin of the American Meteorological Society* 88.6 (2007): 913–28. Web. 12 Nov. 2012.

Pielke, Roger, Sr., et al. "Problems in Evaluating Regional and Local Trends in Temperature: An Example from Eastern Colorado, USA." *International Journal of Climatology* 22 (2002): 42–434. Web. 8 Oct. 2012.

Pielke, Roger, Sr., et al. "Unresolved Issues with the Assessment of Multidecadal Global Land Surface Temperature Trends." *Journal of Geophysical Research* 112 (2007): 1–26. Web. 30 Sept. 2012.

"Pilot Groups." *London 21.org*. London 21, n.d. Web. 4 June 2013.

Pincus, Walter. "Radiation Monitors Installed to Check Exposure." *Washington Post* 1 Apr. 1979: A1+. Microfilm.

Porter, Russell. "City Evacuation Plan: 3 Governors and Mayor Weigh Plans to Meet H-Bomb Attack." *New York Times* 12 Mar. 1955: 1+. *ProQuest Historical Newspapers*. Web. 11 June 2012.

Porter, Theodore. *Trust in Numbers*. Princeton: Princeton UP, 1995. Print.

Potts, Gareth. *Regeneration in Deptford London*. *UrbanBuzz*. UrbanBuzz, Sept. 2008. Web. 3 May 2013. PDF file.

Prelli, Lawrence. "The Rhetorical Construction of Scientific Ethos." *Landmark Essays on Rhetoric of Science*. Randy Allen Harris, ed. Mahwah: Lawrence Erlbaum, 1997. Print.

President's Commission on the Accident at Three Mile Island. *Report of the President's Commission on the Accident at Three Mile Island*. Washington, DC: GPO, 1979. Print.

Prichep, Deena. "Oregon Advertising Studio Tracks Fukushima Radiation." *Voanews.com*. Voice of America, 5 Jul. 2011. Web. 2 Jan. 2012.

"Published Papers." *Zooniverse.org*. Zooniverse, 1 Dec. 2012. Web. 22 May 2014.

Radiological Defense. 1961. Film. US Office of Civil Defense. *Archive.org*. Prelinger Archives, Web. 13 June 2012.

RDTN Radiation Map of Japan. Map. *Rdtn.org*. RDTN, 24 Mar. 2011. Web. 2 July 2013.

Revkin, Andrew. "On Weather Stations and Climate Trends." *Nytimes.com*. New York Times, 28 Jan. 2010. Web. 30 Sep. 2012.

"Roger A. Pielke" *Wikipedia.org*. Wikipedia, 15 Dec. 2004. Web. 28 Sept. 2012.

Ropeik, David. "Risk Reporting 101." *Columbia Journalism Review* 11 Mar. 2011. Web. 27 June 2011.

Rosenfeld, Stephen. "A Blow to Nuclear Arrogance." Editorial. *Washington Post* 2 May 1986: A19. Microfilm.

Rowland, Katherine. "Citizen Science Goes 'Extreme.'" *Nature.com*. Nature, 17 Feb. 2012. Web. 20 Feb. 2012.

"Runaway Reactor." *Time* 13 Jan. 1961: 18–19. Print.

Saenz, Aaron. "Japan's Nuclear Woes Give Rise to Crowd-sourced Radiation Maps in Asia and US." *Singularityhub.com*. Singularity Hub, 24 Mar. 2011. Web. 15 Dec. 2011.

Safecast Map. Map. *Safecast.com*. Safecast, 10 Aug. 2011. Web. 14 Aug. 2012.

Safire, William. "The Fallout's Fallout." Editorial. *New York Times* 5 May 1986: A19. Microfilm.

Sauer, Beverly. *Rhetoric of Risk: Technical Documentation in Hazardous Environments*. Mahwah: Lawrence Erlbaum, 2003. Print.

Schmemann, Serge. "Soviet Announces Nuclear Accident at Electric Plant." *New York Times* 29 Apr. 1986: A1+. *ProQuest Historical Newspapers*. Web. 23 Jul. 2011.

———. "Soviet Secrecy." *New York Times* 1 May 1986: A1+. *ProQuest Historical Newspapers*. Web. 22 May 2012.

Shimbun, Yomiuri. "Melt-through at Fukushima? Govt. Report to IAEA Suggests Situation Worse than Meltdown." *Daily Yomiuri Online*. The Daily Yomiuri, 8 June 2011. Web. 8 June 2011.

Sieber, Renee. "Public Participation Geographic Information Systems: A Literature Review and Framework." *Annals of the Association of American Geographers* 96.3 (2006): 491–507. Web. 20 May 2013.

Silvertown, Jonathan. "A New Dawn for Citizen Science." *Trends in Ecology and Evolution* 24.9 (2009): 467–71. Web. 26 Sept. 2012.

Simmons, Michelle. *Participation and Power: Civic Discourse in Environmental Policy Decisions*. New York: SUNY P, 2007. Print.

SIMPLIFi Solutions "Recycling Firm Fined over £190,000 for Causing Years of Nuisance." *Prosecutions Newsletter*, January 2012. Web. 30 May 2013.

Soundings. *The Wharves, Deptford Statement of Community Involvement*. Thewharvesdeptford.com. Soundings, 2010. Web. 16 Mar. 2012. PDF file.

Steigerwald, Bill. "Helping Along Global Warming." *Triblive.com*. Pittsburg Tribune-Review, 17 June 2007. Web. 28 Sep. 2012.

Stewart, Paul. "The Value of Christmas Bird Counts." *The Wilson Bulletin* 66.3 (1954): 184–95. Web. 12 Dec. 2012.

Stinson, John. "Historical Note." *National Audubon Society Records, 1883–1991*. New York: New York Public Library, 1994. Web. 10 Dec. 2012. PDF file.

Suter, David. "Radiation Plume from Three Mile Island." Map. *Harper's* Oct. 1979: 16. Print.

Taylor, James. "Meteorologist Documents Warming Bias in U.S. Temperature Stations." *Heartland.org*. Heartland Institute, 1 Nov. 2007. Web. 30 Sep. 2012.

"Thames Gateway." *Wikipedia.org*. Wikipedia, 3 Apr. 2013. Web. 4 June 2013.

Travierso, Michele. "Tech in Troubled Times: Website Crowdsources Japan Radiation Levels." *Newsfeed.time.com*. Time, 21 Mar. 2011. Web. 5 Jan. 2011.

Tuchinsky, Evan. "Watts, Me Worry?" *Newsreview.com*. Reno News & Review, 6 Dec. 2007. Web. 30 Sep. 2012.

"2008 Presidential Election." *Politico.com*. Politico, n.d. Web. 28 Nov. 2012.

"2012 Presidential Election." *Politico.com*. Politico, 29 Nov. 2012. Web. 28 Nov. 2012.

United States. Department of Defense. *Fallout Protection: What to Know and Do About a Nuclear Attack*. Washington, DC: GPO, 1961. Print.

———. Patent Office and Smithsonian Institution. *Results of Meteorological Observations under the Direction of the United States Patent Office and the Smithsonian Institution from the Year 1854 to 1859, Inclusive*. Washington, DC: GPO, 1864. *Google Book Search*. 11 Dec. 2012. PDF file.

Von Hippel, Frank, and Thomas Cochran. "Estimating Long Term Health Effects." *Bulletin of the Atomic Scientist* 43.1 (1986): 18–24. Print.

Vose, Russell, et al. "Comments on 'Microclimate Exposures of Surface-based Weather Stations.'" *Bulletin of the American Meteorological Society* 86.4 (2005): 497–504. Web. 9 Oct. 2012.

Waddell, Craig. "Saving the Great Lakes: Public Participation in Environmental Policy." *Green Culture*. Eds. Carl Herndl and Stuart Brown. Madison: U of Wisconsin P, 1996. 141–65. Print.

Walker, Kenneth, and Lynda Walsh. "'No One May Yet Know What the Ultimate Consequences May Be': How Rachel Carson Transformed Scientific Uncertainty into a Site For Public Participation in *Silent Spring*." *Journal of Business and Technical Communication* 26.3 (2011): 3–34. Web. 13 July 2013.

Walton, Douglas. *Appeal to Popular Opinion*. University Park: The Pennsylvania State UP, 1999. Print.

Warnick, Barbara, and David Heinemann. *Rhetoric Online: The Politics of New Media*. 2nd ed. New York: Peter Lang, 2012. Print.

Watanabe, Takeshi. "Tokyo Hacker Space Gets the Data." *Majiroxnews.com*. Majirox News, 20 May 2011. Web. 5 Jan. 2012.

Watts, Anthony. "Another Milestone—200 Volunteers." *Wattsupwiththat.org*. Anthony Watts, 3 Aug. 2007. Web. 13 Nov. 2012.

———. "Day 2 at NCDC and Press Release: NOAA to Modernize USHCN." *Wattsupwiththat.com*. Anthony Watts, 24 Apr. 2008. Web. 23 Oct. 2012.

———. *A Hands-on Study of Station Siting Issues for United Stated Historical Climatology Network Stations*. *SurfaceStations.org*. Surface Stations, 29 Aug. 2007. Web. 11 Nov. 2012. Power Point.

———. "How to do a USHCN, GSHCN, or GISS Weather Station Site Survey." *Surfacestations.org*. Anthony Watts, 16 June 2007. Web. 10 Oct. 2012.

————. *Is the U.S. Surface Temperature Record Reliable?* Chicago: Heartland Institute, 2009. Web. 26 Sept. 2012.

————. "NOAA and NCDC Restore Data Access." *Wattsupwiththat.com.* Anthony Watts, 7 July 2007. Web. 28 Sep. 2012.

————. "NOAA/NCDC Throw a Roadblock in My Way." *Wattsupwiththat.com.* Anthony Watts, 30 June 2007. Web. 28 Sep. 2012.

————. "Re: A New Paper on the Differences between Recent Proxy Temperature and In-situ Near-surface Air Temperatures." *Climate Science: Roger Pielke Sr.* Roger Pielke Sr., 4 May 2007. Web. 12 Nov. 2012.

————. "Road Trip Update: Day 1 at NCDC." *Wattsupwiththat.com.* Anthony Watts, 23 Apr. 2008. Web. 23 Oct. 2012.

————. "Site Survey: Weather Station of Climate Record at CSUC." *Wattsupwiththat .com.* Anthony Watts, 9 May 2007. Web. 11 Nov. 2012.

————. "Standards for Weather Stations Siting Using the New CRN." *Wattsupwiththat .com.* Anthony Watts, 3 July 2007. Web. 10 Nov. 2012.

————. *Surfacestations.org.* Anthony Watts, 4 June 2007. Web. 29 Sep. 2012.

————. "Surfacestations.org Is Ready and Your Assistance Is Needed!" *Climate Science: Roger Pielke Sr.* Roger Pielke Sr., 5 June 2007. Web. 28 Sep. 2012.

————. "2006 Hottest Year on Record—So What? Part 1." *Wattsupwiththat.com.* Anthony Watts, 10 Jan. 2007. Web. 11 Nov. 2012.

"What Is Tokyo Hackerspace?" *Tokyohackerspace.org.* Tokyo Hackerspace, n.d. Web. 4 Jan. 2012.

Whitaker, Coleen. *This Is Where We Live! The Community Mapping Action Pack. London21.org.* London 21, 2008. Web. 5 May 2013. PDF file.

White, Charles. "New Website Crowdsources Japan Radiation Data." *Mashable.com.* Mashable Social Media, 20 Mar. 2011. Web. 17 Dec. 2011.

Will, George. "Mendacity as Usual." Editorial. *Washington Post.* 4 May 1986: C8. Microfilm.

World Health Organization. *Guidelines for Community Noise.* Ed. Brigitta Berglund, Thomas Lindvall, and Dietrich Schwela. Geneva: World Health Organization, 1999. Web. 26 May 2013.

————. "Occupational and Community Noise." *Euro.who.int.* World Health Organization, Feb. 2001. Web. 14 June 2013.

World Meteorological Organization. *Guide to Meteorological Instruments and Methods of Observation.* 6th ed. Geneva: WMO, 1996. Print

World Nuclear Association. "Fukushima Accident 2011." *World-nuclear.org.* N.d. Web. 8 June 2011.

Yanch, Jacqueline. "Background Information about Radiation and What It Does." *Safecast.org.* Safecast, 7 Apr. 2011. Web. 6 July 2012.

Zhang, Hiayan. Invited Comment. "When Crowdsourcing Data Meets Nuclear Power." By Alexis Madrigal. *Theatlantic.com.* The Atlantic, 24 Mar. 2011. Web. 17 Dec. 2011.

Zuckerman, Ethan. "Mohammed Nanabhay and Joi Ito at Center for Civic Media." *Ethanzuckerman.com.* 27 Jun. 2011. Blog. 7 Jan. 2012.

Index

acoustical measurement, 130–31, 149–54;
   by the borough on Pepys Estate, 149;
   citizen science protocols for, 131, 175; by
   citizen scientists on Pepys estate, 138–
   40, 141–43; and policy action, 150–52
acoustic consultant, 149, 151, 154
acoustic survey, 150–51
ad numerum, 89–90, 93
ad populum, 89, 160, 185n5
AIDS, 1, 4, 9, 68, 73, 164–65, 168
Akiba (Christopher Wang), 57–58, 81, 86–
   87, 185n19
Alexander, Heidi, 139, 151–54, 189n28
Alvarez, Marcelino, 53–54, 57–58, 74–75,
   79–80, 90
American Ornithologists' Union (AOU),
   18–19
analogy, 45–46, 78
arête (virtuousness), 71, 73, 76, 87, 91–92,
   156; epistemological objectivity, 73; general discussion of, 71; nontechnical, 73,
   156; technical, 73, 87, 91–92
Aristotle, 71–72, 124, 133
Atomic Energy Commission (AEC), 35, 92
Audubon Society/Societies, 17–23, 26, 183n5

benivolentia, 72
bGeigie, 59–60, 165
Bird Lore (journal), 18, 20, 183n6
Bogost, Ian, 11, 67
Bonner, Sean, 56–58, 60, 62–64, 74, 76,
   82, 88, 90
Bonney, Rick, 5, 169

bull's-eye overlay, 32–37, 44, 47, 51, 62–63,
   185n13; consequences of use of, 33–37;
   history of use 32–33. *See also* radiation
   mapping

Center for the Advancement of Informal
   Science Education (CAISE), 7, 167–68
Chapman, Frank, 17–20
Chernobyl, 9, 28–29, 31, 37, 50, 64, 164; accident account, 37; Soviet cover up of,
   38–40. See also *New York Times*; *Washington Post*
Christmas bird census. *See* Christmas Bird
   Count
Christmas Bird Count (CBC), 17–22, 23, 26,
   183n1; conflict over count methods, 20–
   22; datacollection methods, 19; as education, 19; as entertainment, 19–20; exigence for, 18; goal of, 17–18; scientific
   fixes for count methods, 21–22; success of, 20
Cicero, 72
citizen science, 8–11, 117–18, 165, 172–73;
   benefits and challenges of, 4–7, 169–
   72; and crowdsourcing, 79; definition,
   2–3; grassroots, 26, 28–29, 53, 65, 168;
   past and present, 12, 22–27; and public
   policy, 10, 128–29, 133, 136–38, 159; as a
   resource for invention, 128–29; types of,
   95–96. *See also* acoustical measurement;
   Internet; *logos*; Pepys Estate
citizen scientists, 2, 4, 10–11, 23–24, 44, 77–
   80, 84, 98, 105, 108, 121, 125–26, 128,